Springer Tracts in Mechanical Engineering

Springer Tracts in Mechanical Engineering (STME) publishes the latest developments in Mechanical Engineering - quickly, informally and with high quality. The intent is to cover all the main branches of mechanical engineering, both theoretical and applied, including:

- Engineering Design
- Machinery and Machine Elements
- Mechanical Structures and Stress Analysis
- Automotive Engineering
- Engine Technology
- Aerospace Technology and Astronautics
- Nanotechnology and Microengineering
- Control, Robotics, Mechatronics
- MEMS
- Theoretical and Applied Mechanics
- Dynamical Systems, Control
- Fluids Mechanics
- Engineering Thermodynamics, Heat and Mass Transfer
- Manufacturing
- Precision Engineering, Instrumentation, Measurement
- Materials Engineering
- Tribology and Surface Technology

Within the scope of the series are monographs, professional books or graduate textbooks, edited volumes as well as outstanding PhD theses and books purposely devoted to support education in mechanical engineering at graduate and post-graduate levels.

Indexed by SCOPUS, zbMATH, SCImago.

Please check our Lecture Notes in Mechanical Engineering at http://www.springer.com/series/11236 if you are interested in conference proceedings.

To submit a proposal or for further inquiries, please contact the Springer Editor **in your country**:

Dr. Mengchu Huang (China)
Email: mengchu.Huang@springer.com
Priya Vyas (India)
Email: priya.vyas@springer.com
Dr. Leontina Di Cecco (All other countries)
Email: leontina.dicecco@springer.com

More information about this series at http://www.springer.com/series/11693

Rajiv Kumar Sharma

Quality Management Practices in MSME Sectors

Springer

Rajiv Kumar Sharma
Department of Mechanical Engineering
NIT Hamirpur
Hamirpur, India

ISSN 2195-9862 ISSN 2195-9870 (electronic)
Springer Tracts in Mechanical Engineering
ISBN 978-981-15-9511-0 ISBN 978-981-15-9512-7 (eBook)
https://doi.org/10.1007/978-981-15-9512-7

This Springer imprint is published by the registered company Springer Nature Singapore Pte Ltd.
The registered company address is: 152 Beach Road, #21-01/04 Gateway East, Singapore 189721, Singapore

Preface

According to the Ministry of MSMEs (India), 95% of industries are working in small-scale sector with 40% value addition in the manufacturing sector and 8% contribution to the gross domestic product (GDP). After agriculture sector, micro small and medium enterprise (MSME) sector is the second largest employer as it provides employment to an estimated 117 million people spread over 51 million enterprises (Reserve Bank of India, 2017) report. The MSMEs produce more than 8000 different products ranging from the conventional products to complicated products (in terms of high-technology intensive products). It is estimated that in terms of value, the sector accounts for 45% of the manufacturing output and 40% of the total export (Ministry of MSME's, 2017). From the current economic forecast, it is inferred that Indian economy is growing at an average of 7% per annum and is likely to become a $5 trillion economy by the year 2025. Hence, the growth of the MSMEs sector becomes indispensable as it is regarded as the backbone of India. India has all resources and possesses the necessary skills, but it still lags far behind developed nations. The foremost challenge for MSMEs is to offer novel and custom-made products using the best technology.

Over the last three decades, MSME sector has emerged as a highly vibrant and dynamic sector of the Indian economy. This sector accounts for a significant proportion of industrial enterprises, employment, national income and export in developed as well as in developing countries. Quality has been widely recognized as one of the most important discipline/strategies or competitive priority for an organizational development by the business houses. Quality Management (QM) practices have been widely used by large business houses such as Hero, Tata Motors, Maruti Suzuki, Motorola, Ford and Honda, IOCL etc, for competitive positioning but their adoption and practices in MSME's sector are not adequate, which calls for necessary attention by the researchers in this field. There is no doubt that Indian MSMEs are passing through a transitional period of high-market competition with escalating demands of consumers for getting better product and service. For the survival of MSMEs in this ever-expanding marketplace and to achieve customer satisfaction and improvement in productivity, the emergence of quality plays a vital role. Taking note of this, some MSMEs have begun to adopt QM practices as an approach to fulfill these objectives. However, there is a strong need to understand the importance of these practices for MSMEs, so as they can improve their current business climate through improved quality of product and service, and ensure their long-term survival. Therefore, this study has multiple objectives of investigating the relationship between quality management practices and competitive positioning in MSMEs. The present study covered MSMEs in Himachal Pradesh, India.

The significance of quality for company's performance is widely established in business literature and company practices. In order to improve business performance and achieve competitiveness, numerous approaches have been pursued in literature, most remarkably and suggested approach is the concept of quality management, which has its roots since early 1900s, i.e., preindustrial revolution. In recent times, total quality management (TQM) has been a burning topic in business literature. It is a holistic approach that seeks managing quality; it required the development of quality approach and a frame or structure for its successful implementation. It augments the conventional way of doing a business based on the principle of a continual improvement of the organizational practices. More specifically, it is the application of sound management principles, quantitative methods, and management of human resources to develop all the process within an organization and meet customer requirements. It has been widely acknowledged as an essential management practice, which can play a key role in making companies to become more competitive in global economy. To succeed in business companies shall adopt quality management practices as a source of major organizational change that requires a change in the organizational culture, processes, and strategies or priorities.

In the light of this situation and global competitiveness, there is need for a deeper investigation of relationship between QM practices and MSMEs performance. It is to fill this gap, the present volume embodies a humble effort to convey such a holistic and comprehensive view of quality to acquaint the readers with the following topics of importance in form of book chapters

- Significance and importance of adopting of quality management practices among MSMEs.
- Collection of useful data pertaining to awareness regarding quality management.
- Categorizing ISO certified MSMEs on the basis of pollution control board norms.
- Assessment of the willingness of SMEs to develop their systems of quality and enhance their level of quality standards such as ISO 9000 in various MSME sectors.
- Showcase the applicability of various quality tools in MSMEs and stages in company where quality management practices are applied.
- To develop a measure based on Porter's Diamond model indicators for competitive positioning in MSMEs.
- Two case studies to summarize the benefits obtained from implementing various quality practices in SMEs.

The present volume is not intended to be a textbook on Quality or Quality Management. It is meant to express an extensive understanding of Quality system prevailing in different MSME sectors.

Chapter 1 presents the introduction to Indian micro small and medium enterprises (MSMEs), importance of quality in business processes, brief account of history of evolution of quality management, statistics required for quality management, data collection and measurement scales, measures of central tendency and summary of literature studies. Chapter 2 of the book provides the extensive discussion on tools and techniques used for the introduction of quality management practices. Chapter 3 introduces the overall research design and methodology adopted along with the design of survey instrument. Chapter 4 details the need for introducing quality management practices. Chapter 5 presents the discussion to the nature of quality management tools applied in MSMEs, stages where they applied and year of adoption of these quality practices in MSMEs. Chapter 6 presents the application of Porter's diamond model variables for assessing competitiveness among various MSME sectors. Chapter 7 evaluates the company's experience after implementing quality management practices, Chapter 8 details the analysis related to various critical success factors, which can help MSMEs in adoption of quality practices. Chapter 9 presents the comparative analysis of competitive positioning among MSME sectors.

The materials presented are mostly reflections of the author's own realizations and writings, though some materials contained in some books or reports or documents have also been used with due reference. The book also enlightens the readers with two case studies related to quality management practices.

Hamirpur, India — Rajiv Kumar Sharma

Acknowledgements

This book has been an outcome of the research conducted at NIT Hamirpur and I am indebted to persons from the organization that directly or indirectly helped during the preparation of the manuscript. It is my pleasure to acknowledge their help. I would like to thank my beloved parents (Shri. Jyoti Prashad sharma and Smt. Chanchal Sharma) and my wife Pooja Sharma, my daughter Anandita Sharma for their encouragement and support for completing this project.

I also wish to thank Priya Vyas, Associate Editor—Applied Sciences & Engineering, Springer—a part of Springer Nature, for her enthusiasm, patience, and support since the beginning of the book, as well as Maniarasan (Mr.) Project Coordinator, Books Production (Springer), and his staff for their meticulous effort regarding production.

Contents

1 Quality Management and MSMEs 1
1.1 Introduction 2
1.2 MSMEs Classification 2
1.3 Importance of Quality in Business Processes 5
1.4 Evolution of Quality Management 6
1.5 Statistics for Quality Management 7
1.5.1 Data Collection 9
1.5.2 Measurement Scales 10
1.5.3 Measures of Central Tendency 11
1.6 Literature Review on National and International Studies 14
1.7 Concluding Remarks 15
References 23

2 Quality Management Tools and Standards 25
2.1 Introduction 25
2.2 Quality Management Tools 26
2.2.1 Cause and Effect/Fishbone/Ishikawa Diagram 27
2.2.2 Check Sheet 28
2.2.3 Control Charts 30
2.2.4 Process Flow Chart 34
2.2.5 Histogram 36
2.2.6 Pareto Diagrams 37
2.2.7 Scatter Plot 39
2.2.8 Customer Relationship Management 41
2.2.9 Enterprises Resource Planning 43
2.2.10 Kaizen 44
2.2.11 FMEA (Failure Mode and Effects Analysis) 45
2.2.12 Kanban 48
2.2.13 Poka-Yoke 49
2.2.14 Quality Function Deployment 50
2.2.15 Supply Chain Management 52
2.2.16 Six Sigma 54

2.2.17 5S and House Keeping ... 56
2.3 Quality System Standards ... 56
2.3.1 ISO 9000 and Other Derivatives ... 58
2.4 Action Plan for Implementing ISO 9001:2008 Standards ... 59
2.5 Concluding Remarks ... 59
References ... 61

3 Quality Management in MSME Sectors ... 63
3.1 Introduction ... 63
3.2 Problem Definition ... 64
3.2.1 Study Objectives ... 64
3.2.2 Research Methodology ... 65
3.2.3 Sample Size Determination ... 67
3.2.4 Design of Pilot Instrument ... 70
3.2.5 Pretesting and Validation of Questionnaire ... 73
3.2.6 Respondents and Their Profile ... 73
3.3 Concluding Remarks ... 73
Annexure 1: General Information ... 74
Annexure 2: Information on Need of Quality Management Practices and Quality Standard Certification ... 75
Annexure 3: Information on Quality Management Practices ... 76
Annexure 4: Information on Competitiveness ... 77
Annexure 5: Information on Benefits After Implementing Quality Management Practices ... 78
Annexure 6: Information on Critical Success Factors for Adoption of Quality Management Practices ... 79
References ... 80

4 Need of Quality Management Practices ... 81
4.1 Introduction ... 81
4.2 Data Collection ... 81
4.3 Respondents Profile ... 82
4.4 Data Analysis ... 82
4.4.1 Response Profile ... 82
4.4.2 Need of Quality Management Practices ... 84
4.5 Concluding Remarks ... 92
References ... 94

5 QMS Implementation Stages ... 95
5.1 Introduction ... 95
5.2 Sector of Operation ... 95
5.3 Year-wise Adoption of Quality Practices/Programs ... 96
5.4 Stages in Company Where QM Practices Applied ... 98
5.5 Nature of Quality Management Tools Being Applied ... 99
5.6 Concluding Remarks ... 100

6 Information on Competitiveness 103
6.1 Introduction 103
6.2 Factor Conditions 103
6.2.1 Presence of Natural Resource 105
6.2.2 Role of Physical Resources 105
6.2.3 Role of Unskilled Labor 106
6.2.4 Role Highly Educated Personnel 106
6.2.5 Role of Advanced Production and Process Technology 107
6.2.6 Role of IPR and Patents 107
6.2.7 Role of Communication Infrastructure 108
6.3 Demand Conditions 109
6.3.1 Role of Market Size 109
6.3.2 Role of Pattern of Growth 109
6.3.3 Role of Sophistication 110
6.3.4 Role of Effect of New Investment in the Region 110
6.4 Related and Supported Industries 111
6.4.1 Role of Technology Upgradation 111
6.4.2 Role of Information Flow 112
6.4.3 Role of Financial Management 112
6.4.4 Role of R&D Investments 113
6.4.5 Role of Supplier and Distribution Channel 113
6.4.6 Role of Importance of Marketing 114
6.5 Firm Structure, Strategy, and Rivalry 114
6.5.1 Role of Boosted Innovation 114
6.5.2 Role of Internal Structure 115
6.5.3 Role of Geographic Concentration 115
6.6 Government and Culture 116
6.6.1 Role of Financial Support System 116
6.6.2 Role of Environmental Regulations 116
6.6.3 Role of Impact of National Culture 117
6.6.4 Role of Business Climate 118
6.7 Concluding Remarks 118

7 Information on Benefits After Implementing QM Practices 121
7.1 Introduction 121
7.2 Benefits of QM Implementation 122
7.2.1 Employee Training 122
7.2.2 Employee Participation 122
7.2.3 Effect of Employee Morale 127
7.2.4 Effect on Competitive Spirit 127
7.2.5 Quality Awards 129
7.2.6 Recruitment and Retaining 129
7.2.7 Warranty Expenditure 129
7.2.8 Customer Complaints 130

7.2.9 Manufacturing Lead Time 130
7.2.10 Energy Consumption 130
7.2.11 Maintenance Cost 131
7.2.12 Rework and Scrap 131
7.2.13 Sales Volume 131
7.2.14 Company Profit 132
7.2.15 Customers Relations 132
7.2.16 Supplier Relationship 132
7.3 Structural Equation Model for Relationship Among Variables 133
7.4 Concluding Remarks 138
References 139

8 Critical Success Factors in Adoption of QM Practices 141
8.1 Introduction 141
8.2 Strategic Factors 141
8.2.1 Top Management Commitment 142
8.2.2 Quality Culture 143
8.2.3 Quality Systems 143
8.2.4 Quality Awards 144
8.2.5 Continuous Improvement and Innovation 144
8.2.6 Benchmarking 145
8.2.7 Technical Expertise 145
8.2.8 Role of Market Expertise 146
8.2.9 Role of Financial Access 147
8.3 Tactical Factors 147
8.3.1 Role of Employee Encouragement 147
8.3.2 Role of Employee Involvement 147
8.3.3 Training and Education 148
8.3.4 Teamwork 149
8.3.5 Information and Analysis 149
8.3.6 Supplier Management 150
8.3.7 Communication 151
8.4 Operational Factors 151
8.4.1 Product and Service Design 151
8.4.2 Process Management 151
8.4.3 Customer Focus 152
8.4.4 Human Resource Management 153
8.4.5 Industrial Relations 153
8.4.6 Employee Rewards and Incentives 154
8.5 Concluding Remarks 154

9 Comparative Analysis of Competitive Positioning Among MSMEs Sectors 157
9.1 Introduction 157
9.2 Michael Porter's Competitive Approach 158
9.2.1 Factor Conditions 158

9.2.2 Demand Conditions 159
9.2.3 Related and Supporting Industries 159
9.2.4 Firm Strategy, Structure and Rivalry 160
9.2.5 Government and Chance Events 160
9.3 Current Competitive Positioning of MSME Sectors 161
9.3.1 Pharmaceutical Sector 162
9.3.2 Electronics Sector 164
9.3.3 Automobile Sector 165
9.3.4 Food Sector 165
9.3.5 Textile Sector 166
9.4 Concluding Remarks 167
References 169

Appendix A: Case Study 1 171

Appendix B: Case Study 2 189

About the Author

Dr. Rajiv Kumar Sharma is presently working as an Associate Professor and has also served as the Head of Department of Mechanical Engineering NIT Hamirpur (H.P). He obtained his B.Tech. (Hons) from SLIET Longowal (CFTI), Punjab, M.Tech. (Hons) in Industrial and Production Engineering with Gold Medal from Thapar University Patiala and doctoral degree (Ph.D.) from Indian Institute of Technology; IIT Roorkee-247667.

Till so far he has guided 05 Ph.D. theses, 25 M.Tech. Thesis and more than 50 UG projects in various areas of engineering. He has research and teaching experience of more than 15 years and about 100 publications in international journals of repute. He is also on Editorial Board of many reputed journals such as International Journal of Quality and Reliability Management, TQM Journal, Emerald Publishers; International Journal of Strategic Business Alliance, Inderscience Publishers, Advances in Production Engineering & Management, Production Engineering Institute (PEI), University of Maribor, Maribor, Slovenia European Union (EU). He has completed sponsored research projects from UGC and DST. His research interests include Supply chain and Logistics, Flexible manufacturing systems, statistical quality control, system reliability and maintenance, quality engineering, six sigma, lean manufacturing and Industry 4.0.

Chapter 1
Quality Management and MSMEs

Highlights

- *Micro, small and medium enterprises (MSMEs) are regarded as "life blood of modern economies" as one of the main dynamic forces of economic development, which stimulates private ownership and entrepreneurial skills.*
- *According to the Ministry of Small and Medium Enterprises, India, 95% of industrial units are in small-scale sector with 40% value addition in the manufacturing sector and 8% contribution to the Indian gross domestic product (GDP).*
- *Indian MSMEs are passing through a transitional period of high market competition with increasing demands of customers for getting improved product and service.*
- *For the survival of MSMEs in this ever-expanding marketplace, customer satisfaction, enhancement in productivity, changing organizational culture and globalization of world trade, the implementation and practice of quality management plays fundamental role in their competitive positioning.*
- *The chapter presents details to*
  - *Importance of MSMEs in Indian economy, their classification and importance of quality.*
  - *Evolution of quality management.*
  - *Statistics for quality management.*
  - *Summary of literature studies.*

R .K. Sharma, *Quality Management Practices in MSME Sectors*,
Springer Tracts in Mechanical Engineering,
https://doi.org/10.1007/978-981-15-9512-7_1

1.1 Introduction

Considered to be the *"life blood of modern economies"* and regarded as *"one of the main driving forces"* of economic development, MSMEs hold opposing views from large business houses in various aspects such as management style, production operations, capital availability, purchase and procurement procedures, inventory and quality control systems, and negotiating power. MSMEs are quite flexible and can adjust quickly to varying market demand conditions and supply situations. They generate employment opportunities and make noteworthy contribution to exports and foreign trade stimulating private ownership and entrepreneurial skills to a great extent. The accomplishments of large organizations in producing or improving the quality of products or services mostly depend on the quality of products or components supplied to them by the supplier firms. Usually, the component suppliers are small- and medium-sized business organizations. Therefore, MSMEs face substantial pressure to achieve ISO 9000 quality management certification, a widely known and popular quality system certification standard.

The foremost challenge for MSMEs is to offer novel and custom-made products using the best technology. For success in businesses, MSMEs in all sectors must build up effective strategies in terms of cost, quality and services at shortest possible time (Porter 1990) to provide better values to the customers. There is no doubt that MSMEs have made noteworthy contribution toward technical and scientific growth, with increase in export potential. Various key sectors such as food and beverage processing, agriculture, engineering, electrical, electronics, electromedical equipment, textiles and garments, chemicals and pharmaceuticals, processed food, textiles, information technology, gems and jewelry, chemicals, ayurvedic products, leather and leather goods have recognized their domestic as well as global presence.

1.2 MSMEs Classification

MSMEs sector has come into sight as a highly energetic and self-motivated sector of the Indian economy over the last three decades. According to Ministry of MSMEs (India), 95% of industries are working in small-scale sector with 40% value addition in the manufacturing sector and 8% contribution to the gross domestic product (GDP). After agriculture sector, MSME sector is the second-largest employer as it provides employment to an estimated 117 million persons spread over 51 million enterprises (Reserve Bank of India 2017 report). The MSMEs produce more than 8000 different products ranging from the conventional products to complicated products (in terms of high-technology intensive products). It is estimated that in terms of value, the sector accounts for 45% of the manufacturing output and 40% of the total export (Ministry of MSME's 2017). From the current economic forecast, it is inferred that Indian economy is growing at an average of 7% per annum and is likely to become a $5 trillion economy by the year 2025. Hence, the growth of the MSMEs sector becomes

Table 1.1 MSMEs classification[a]

Classification	Description
Micro enterprise	The investment in Plant and Machinery or Equipment does not exceed 1 crore rupees and turnover does not exceed 5 crore rupees
Small enterprise	The investment in Plant and Machinery or Equipment does not exceed 10 crore rupees and turnover does not exceed 50 crore rupees
Medium enterprise	The investment in Plant and Machinery or Equipment does not exceed 50 crore rupees and turnover does not exceed 250 crore rupees

[a]*Source* https://www.msme.gov.in

indispensable as it is regarded as the backbone of India. India has all resources and possesses necessary skills, but it still lags far behind developed nations.

According to MSMEs Development Act 2006 (Ministry of Small and Medium Enterprises, India), the enterprises are categorized based on ceiling for plant, machinery or equipments. The details are presented in Table 1.1. MSMEs are complementary to large industries as subsidiary units, which contribute enormously to the socioeconomic growth of the country. In the last two decades, MSMEs working in different sectors have successfully set up activities outside their domestic markets and today to cope up with increased global economic competition they are functioning as an economic growth engine with increased employment opportunities.

Various nations in developing countries have recognized the worth of MSME enterprises, which are seen as the engine of the economic growth, but still they face numerous challenges such dearth of finance, unavailability of specialized talent, requirement of modern technologies, etc. In present market circumstances, in which customer wants better product and service with lower price, organizational culture and quality play very important role for the continued existence of MSMEs. The companies with technology advantage and labor-intensive production capabilities are usually attending high overall performance. According to Khanna et al. (2011), Bhaumik et al. (2015), Indian manufacturing organizations are in urgent need of new strategies, approaches and techniques for meeting their competitive talent. However, in spite of good projections, the Indian MSMEs are facing notable barriers such as deficiency of timely credits, sourcing of raw materials at viable price, insufficient infrastructure amenities as well as power, water and road, and lack of experienced manpower for carrying out activities related to manufacturing and services, etc. (Mukherjee 2018). Firms usually face the challenges of accomplishing performance (market capitalization), and globalization (international intensity) to fight efficiently against global business giants. Several firms are trying hard not only to satisfy their customer's needs but also where feasible go beyond to meet them. This is possible through cost cutback, enhancement in product performance, improved customer satisfaction and steady efforts toward world class businesses. In order for businesses to stay alive and grow up in the future, it is crucial that they deliver high-quality goods and services. Those who inculcate quality are the ones who will flourish in the coming century.

As per the National Sample Survey (NSS) 73rd round, conducted by National Sample Survey Office, Ministry of Statistics and Programme Implementation during the period 2015–16, there were 633.88 lakhs unincorporated nonagriculture MSMEs in the country engaged in different economic activities (196.65 lakh in Manufacturing, 0.03 lakh in Non-captive Electricity Generation and Transmission, 230.35 lakh in Trade and 206.85 lakh in Other Services) excluding those MSMEs registered under (a) Sections 2 m (i) and 2 m (ii) of the Factories Act, 1948, (b) Companies Act, 1956 and (c) construction activities falling under Section F of National Industrial Classification (NIC) 2008. Table 1.2 and Fig. 1.1 show the distribution of MSMEs activity' wise.

Table 1.2 Distribution of MSMEs activity wise

Activity category	Estimated number of enterprises (in lakh)			Share (%)
	Rural	Urban	Total	
(1)	(2)	(3)	(4)	(5)
Manufacturing	114.14	82.50	196.65	31
Trade	108.71	121.64	230.35	36
Other services	102.00	104.85	206.85	33
Electricity[a]	0.03	0.01	0.03	0
All	**324.88**	**309.00**	**633.88**	**100**

[a]Non-captive electricity generation and transmission
(Annual Report 2018–2019 MSME, Govt of India)

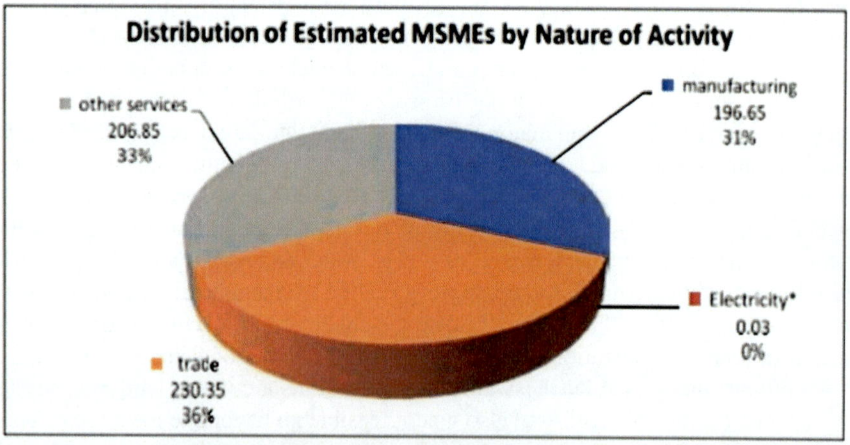

Noncaptive electric power generation, transmission and distribution by units not registered with the Central Electricity Authority (CEA)

Fig. 1.1 Distribution of Estimated MSMEs (Nature of Activity Wise)

1.3 Importance of Quality in Business Processes

Business has tried to define quality in a producer–consumer context, with the following variations. Experts in organizations are seeking to describe quality in the producers' and consumers' perspective, such as:

(a) ISO 9000: The extent to which product's features are successful in fulfilling customer's expectations. This definition is in the context of the consumer.
(b) Six Sigma: The extent to which a process is capable of producing products with minimum variations (i.e., defects per million opportunities DPMO).

The significance of quality for company's performance is widely established in business literature and company practices (Kumar et al. 2009). In order to improve business performance and achieve competitiveness, numerous approaches have been pursued in literature, most remarkably and suggested approach is the concept of quality management (Claver et al. 2003; Talib et al. 2013) which has its roots since early 1900s, i.e., preindustrial revolution. In recent times, total quality management (TQM) has been a burning topic in business literature (Sharma and Kodali 2008). It is a holistic approach that seeks managing quality; it required development of quality approach and a frame or structure for its successful implementation. It augments the conventional way of doing a business (Anupam et al. 2008) based on the principle of a continual improvement of the organizational practices. More specifically, it is the application of sound management principles, quantitative methods, and management of human resources to develop all the process within an organization and meet customer requirements. TQM has been widely acknowledged as an essential management practice, which can play key role in making companies to become more competitive in global economy. To succeed in business companies shall adopt TQM practices as quality management is regarded as a source of major organizational change that requires a change in the organizational culture, processes, and strategies or priorities.

Quality of a process depends upon how efficiently some input is transformed into some output. For this purpose, a process is subdivided into various work items each of which may have a particular intermediate output. The general ways to determine quality of a process or intermediate outputs are defined in terms of internal failures such as (i) defect: it is defined as deviation from the relevant specification, which is not as per the process plan (ii) rework: the work items that can be corrected by repeating some work to eliminate the deviations (iii) Reject or Scrap: the items that cannot be corrected and are treated as rejects. In terms of external failures, the general ways to determine the quality of a process is: (i) lost sales: when not able to meet customer demand, (ii) loss of goodwill, (iii) after sales services and (iv) increase in warranty/guarantee charges, etc. The above measurements apply both to outputs from service operations and to outputs of manufacturing processes.

1.4 Evolution of Quality Management

The primeval Egyptians established a dedication toward quality in building their pyramids. They set tall standards in arts and crafts. The cities, churches, bridge and roads built by Romans motivate us even today. The period 1800–1900 (preindustrial revolution) was called operator quality control period in which production of products and services was primarily confined to single or small group of individuals. The liability of quality was entrusted with a particular individual or small group. Skilled craftsmen/operators controlled their own quality through pride or workmanship. This leads to a sense of achievement for operators that helped to lift their morale and motivate them to attain excellence in their work. Further, the historical evolution of quality practices has taken place in four phases, as depicted in Fig. 1.2

(1) Period (1900–1920)—Quality Inspection, i.e., foreman quality control
(2) Period (1920–1940)—Quality control, i.e., inspection quality control
(3) Period (1940–1960)—Quality assurance, i.e., statistical quality control
(4) Period (1980s and beyond)—Total quality management.

Foreman quality control had its beginning from the factory system that expanded following the industrial revolutions. Due to development of industrial revolution, the notion of mass production, assembly lines, and division or span of labor came to existence. Products were generally made from nonstandardized materials and using nonstandard methods. The operator was not held accountable for the production of the entire products but only for a particular part of it. Inspection is carried out mainly to separate nonconforming quality products from the conforming ones, which were later on, send for rework or sold at discounted price. Or we can say that inspection was mainly carried out to ensure nonconforming products through visual inspection by naked eye or by using some sort of testing or inspection techniques. The foreman on shop floor had several operators working under his supervision and was responsible for maintaining quality in his area.

The second stage of quality evolution started during 1920 owing to increase in production quantity, intricacy in product design, process procedures became complex. With increase in number of workers reporting to a foreman, it became not

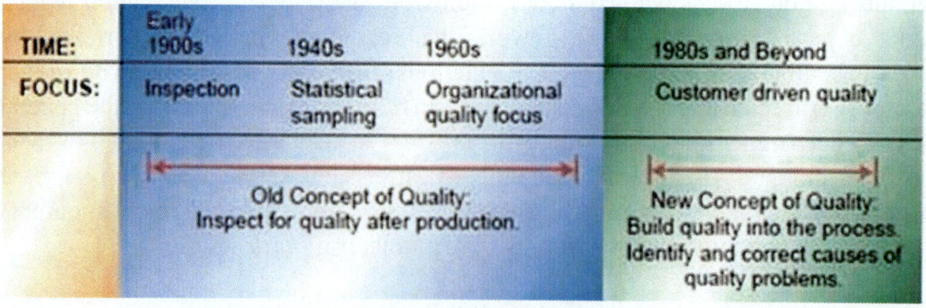

Fig. 1.2 Historical development of quality management

possible for the foreman to keep close control over individual workers. During the period 1920–1940, the specifications, measurement requirements and standardization with respect to products were stated in written form. This helped the inspectors to compare the quality of products against standards. The concept of statically process control came into picture by the development of control charts (Shewhart's), followed by acceptance-sampling plans by Dodge-Roming. This helped the organizations to substitute 100% inspection procedure as it becomes difficult and monotonous to carry out 100% inspection.

The third phase of quality evolution laid emphasis on detection activities, which may help to prevent loss because of poor quality. The aim of the quality assurance function is to develop a system that incessantly reviews the effectiveness of quality philosophy of the organization. In other words, quality assurance works out plans or formulates systematic actions required to assure the performance of the product. It consists of various activities that include systematic audits, failure mode and effect analysis, Taguchi design of experiments and other programs of similar nature. Most of these techniques were designed to avert sudden field failures or eliminate defects. Some other quality management activities such as use of quality costs, quality audits and process development actions were also implemented to give support to quality control.

The fourth phase, started with the word "total" or "holistic" approach to quality. The significant aspect of this phase was the involvement of people across various departments toward controlling the quality of their processes. People began to recognize that each department had important role in making quality product. Further, the concept of zero defects also pictured during this period, its main aim was to improve productivity by doing things right first time. It has become a decisive distinction in today's business where the winning strategy in the business is to achieve customer trustworthiness. Table 1.3 presents the distinct contributions of quality gurus in evolution of quality.

1.5 Statistics for Quality Management

Statistics deals with the collection, classification, analysis and interpretation of numerical facts or data or some information. It is divided into *descriptive statistics* and *inferential statistics*. A study using descriptive statistics is simpler to perform. However, to need evidence about the existence of relationship between variables in an entire population rather than only in the sample, one needs to use inferential statistics.

Descriptive statistics: It illustrates the characteristics of a product or process by making use of the information collected on it. For instance, we wish to describe the test scores out of 100 of class of 30 students. We record all of the test scores and calculated the summary statistics Mean 79.18; range (66.21–96.54) and produce the frequency histogram with horizontal axis representing a range of test score values

Table 1.3 Contributions of quality gurus in the evolution of quality

Name of quality guru	Key contributions
Walter A. Shewhart	• Concept of process inconsistency • Given the notion of statistical process control charts
W. Edwards Deming	• Stressed management's responsibility for quality • Given "14 points" to guide companies in quality improvement
Armand V. Feigenbaum	• Gave the concept of total quality control • Quality defined by customers
Massaki Imai	• Introduced the concept of Kaizen
Philip B. Crosby	• Gave phrase "quality is free" • Introduced concept of zero defects
Shingeo Shingo	• Zero quality control, poka-yoke, JIT related concepts
Kaoru Ishikawa	• Presented cause and effect diagrams • Gave concept of quality circles, companywide quality
G Taguchi	• Focused on product design quality • Gave concept of loss function
Joseph M. Juran	• Provided definition of quality as "fitness for use" • Provided the concept of "cost of quality"

and the vertical axis represents the frequency of scores observed in each test score range (Fig. 1.3). This gives us the details about the scores with ranges. There is no uncertainty surrounding these statistics because the scores for everyone in the class were gathered. However, we can not take these results and extrapolate to a larger population of students.

Inferential statistics: It helps to draw the conclusion from a sample and makes inferences about the larger population from which the sample was drawn. Following steps are used in inferential statistics

1. Defining the population.
2. Drawing a representative sample (random sampling) from the population.
3. Perform analyses that incorporate the sampling error.

The most common methods used in inferential statistics are hypothesis tests, confidence intervals and regression analysis. These inferential methods produce similar summary values as of the descriptive statistics, i.e. the mean and standard deviation.

For example, the effectiveness of a new medication on a particular target group is tested by forming the hypothesis and comparing the outcomes of the proposed sample and then later on generalizes the usefulness of medication on the entire population of

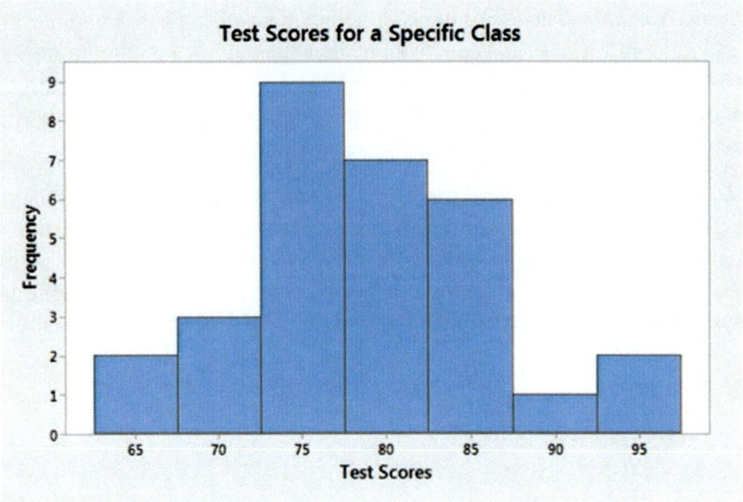

Fig. 1.3 Example of descriptive statistics

patients. For details one may refer to https://statisticsbyjim.com/hypothesis-testing/statistical-hypothesis-testing-overview/.

Confidence intervals incorporate the uncertainty and sample error to create a range of values the actual population value is like to fall within. For example, a confidence interval of [176 186] indicates that we can be confident that the real population mean falls within this range.

1.5.1 Data Collection

In order to exercise control or improve a process, we require data about the process and the data can be collected in several ways. One of the most common ways is through direct observation. In this case, a measurement of the quality characteristic is taken by an observer or automatically by an instrument; for instance, measurements related to size of samples taken by an inspector are the direct observations. Alternatively, in case of indirect observations, the data are collected through questionnaires for instance, if hair dryer company wants to check the field performance of a particular brand of hair dryer than they can assess it by taking a questionnaire survey related to key performance measures. In such case, the data reported by the customers have not been observed by the manufacturer, who exercises no control over the data collection process.

The data on various quality characteristics are expressed by a random variable and are classified as continuous or discrete.

Continuous Variable: The continuous variable takes into account two major points; that it is a random variable and can take on any value within a continuum. Examples of continuous variables are: temperature of an object, speed of a moving object, amount of snow, viscosity of certain resin, thickness of steel plate, particles of sand, etc. Such variables are measurable and have associated numerical values.

Discrete Variable: A discrete variable is a variable whose value is obtained by counting. For instance, the number of train derailments in India, the number of students in a class, the total animals in a zoo, the number of defects in casting, the number of bridges in a state, the number of defective rivets in an assembly, the number of books published and the number of paint blemishes in an automobile, etc.

1.5.2 Measurement Scales

The four distinct measurement scales used for classification of data are (i) nominal, (ii) ordinal, (iii) interval and (iv) ratio scales.

1. *Nominal scale.* This scale is used to refer to the data variables that are simply labels, i.e., conforming and nonconforming or minor, major and critical and are used to identify an attribute of the sample element. Numerical values, even though assigned, are not involved.

2. *Ordinal scale.* The scale of measurement is called ordinal when the data have the properties of nominal data (i.e., labels) as discussed in nominal scale. For instance, customers at a fast-food restaurant are asked to rate the quality of the service according to the given responses, i.e., 1 refers to excellent, 2 refers to good, 3 refers to average, 4 refers to fair and 5 refers to poor response category. There are characterized as ordinal data. Note that a response of 1 does not imply that the service is twice as good as a response of 2. However, we can say that a response 1 is preferable to response 2, and so on.

3. *Interval scale.* When the data have the properties of ordinal data and a fixed unit of measure describes the interval between the observations, the interval scale of measurement is used. Let us assume that we are interested to record the temperature of a furnace. For this, we have taken six readings, i.e., 2000, 2050, 2100, 2150, 2200 and 2250 °F, respectively. These readings are ranked (like ordinal data) in an increasing order of temperature, indicating the lowest temperature, the next lowest and so on.

4. *Ratio scale.* In ratio scale, the data have the properties of interval data. The order and difference between values can be compared. Let us assume that that the weights of five wooden logs are 1.0, 1.2, 1.1, 1.3 and 1.5 kg. Both the order and difference in, the weights of the logs can be compared. We can say that increase in weight from 1.0 to 1.1 is 0.1 kg, which is the same as the increase from 1.2 to 1.3 kg. Also, when

we compare the weights of 1.0 and 1.2 kg, we find a significant ratio. A log weighing 1.2 kg is 10% heavier than one weighing 1.0 kg.

1.5.3 *Measures of Central Tendency*

In this section, measures of central tendency are discussed. These measures are used to derive summarized information from observed values. They specifically tell us somewhat about the position of the observations and the value about which they group and thus help us to make a decision whether the settings of process variables need to be changed or not.

Mean: It is defined as the simple average of the values in a data set. In quality control activities, the mean is the measure that is commonly used to decide whether, the process is moving around the target or not. The sample mean, (represented by X), is obtained by summing up all the observations in a sample and dividing them by the number of observations (n) in that sample. If the ith observation is denoted by X, then the sample mean is obtained from Eq. (1.1) as shown below

$$\bar{X} = \sum_{i=1}^{n} \frac{x_i}{n} \tag{1.1}$$

The populations mean (μ) is obtained by summing all the observations in the population and dividing them by the size of the population (N). It is obtained from Eq. (1.2) as shown below

$$\mu = \sum_{i=1}^{N} \frac{x_i}{N} \tag{1.2}$$

Example 1.1 The management of private bank wishes to reduce the waiting time of the customers. For this, five observations of the customers waiting in the queue in a bank are taken. The waiting times of customers are 2 min, 3 min, 5 min, 1 min and 4 min, respectively. Compute the mean waiting time, which the bank can use to decide whether the waiting time needs to be reduced by employing more service stations.

Solution: The mean waiting time is computed by using Eq. 1.1

$$\bar{X} = \frac{2+3+5+1+4}{5} = \frac{15}{5} = 3\,\text{min}$$

Median: When the observations are ranked, the value that comes in middle is known as the median value. It has the property that 50% of the values are less than or equal to it. If the number of observations (n) is odd, finding the median is easy. First sort the observations in ascending order (or rank the data from smallest observation to largest observation). Then the median will be the observation in rank position $[(n - 1)/2 + 1]$ on this list. If the number of observations (n) is even, then the simple average of the two middle numbers is chosen as the median.

Example 1.2 The inspector made following observations with respect to piston ring diameters (in millimeters): 52.3, 51.9, 52.6, 52.4, 52.4, 52.1, 52.3, 52.0, 52.5 and 52.5. Compute the median value from the observations.

Solution: First we write the observations in ascending order as: 51.9, 52.0, 52.1, 52.3, 52.3, 52.4, 52.4, 52.5, 52.5, 52.6. Then find the observations in the middle, which are 52.3 and 52.4. The median is computed as $(52.3 + 52.4)/2 = 52.35$. From the results, it is observed that median is less influenced by the extreme values in the data set; thus, it is said to be more "robust" than the mean.

Mode: Most frequently occurring value in the data set is called mode.

Example 1.3 The manager in store wants to ascertain the size of circular saws he should stock most to meet customer requirements. The past sales data of 20 circular saws with sizes (in mm) 80, 120, 100, 100, 150, 100, 80, 150, 100, 80,120, 100, 100, 100, 150, 80, 100, 100, 120, 80 are given.

Solution: In this case, the mode is 100 having the highest frequency, i.e., nine occurrences). Hence, the manager may stock the saws with size 100.

Measures of Dispersion

One of the main objectives of quality control is to investigate and reduce the variability of a process. The measures of central tendency discussed above give us hint regarding the location of a data set. They fail to enlighten about the inconsistency of the observations. As a result, analysis is required to understand the measures of dispersion, which shall give information on the variability of the observations around the target value (typically, the mean).

Range: It is the widely used metric used to measure dispersion in quality control, which is stated as the difference between the largest and smallest values in a data set. Range (R) is defined by Eq. (1.3):

$$R = X_L - X_S \tag{1.3}$$

where X_L: the largest observation and X_S: the smallest observation.

Example 1.4 In an airport, the following 10 observations of the time (in minutes) to receive luggage after landing of airplane are recorded: 17, 10, 20, 19, 22, 18, 13, 21, 15, 20. Calculate the range, which may help the management to decide whether the spread is acceptable or not.

Solution: $R = 22 - 10 = 12$ min. The value of range gives an idea of the variability in the observations.

Variance: The variance is used to measure the fluctuation of the observations around the mean. The larger the value, the greater the fluctuation. The population variance σ^2 is given by Eq. (1.4).

$$\sigma^2 = \frac{\sum_{i=1}^{N}(X_i - \mu)^2}{N} \tag{1.4}$$

where μ: the population mean and N: the size of the population.

Standard Deviation: The standard deviation is perhaps the most commonly used measure of dispersal in quality control. Like the variance, it measures the inconsistency of the observations around the target value. It has the same units as the observations and is easier to interpret. The population standard deviation is given by Eq. (1.5),

$$\sigma = \sqrt{\frac{\sum_{i=1}^{N}(X_i - \mu)^2}{N}} \tag{1.5}$$

where μ: the population mean and N: the population size. The sample variance s^2 is given by Eq. (1.6),

$$s^2 = \frac{\sum_{i=1}^{n}(X_i - \overline{X})^2}{n - 1} \tag{1.6}$$

where $\overline{X}$: the sample mean and n: the number of observations in the sample.

Note that the sample variance is simply the sum of the squared deviations of each observation from the sample average divided by the sample size (n) minus one. In most cases, the sample variance is calculated rather than the population variance because the calculation of the later is possible only when every value in the population is known.

Measures of Association

Measures of association: These measures indicate how two or more variables are related to each other. For instance, as one variable increases, how does it influence another variable? Large values of the measures of association indicate a strong relationship between them, and small values indicate a weak relationship between them.

Table 1.4 Procedure to determine PPS score

Number of responses (N)	Number of units scoring					Total point score (TPS)[a]	Percent point score (PPS)
	1 (W_1)	2 (W_2)	3 (W_3)	4 (W_4)	5 (W_5)		$\frac{TPS}{5 * N} 100$

[a]Total Point Score (TPS) = $1 \times W_1 + 2 \times W_2 + 3 \times W_3 + 4 \times W_4 + 5 \times W_5$

Correlation Coefficient

It is defined as a measure of the strength of the linear relationship between two variables in bivariate data. If two variables are denoted by X and Y, then the correlation coefficient "r" of a sample of observations is found using Eq. (1.7),

$$r = \frac{\sum_{i=1}^{n}(X_i - \bar{X})(Y_i - \bar{Y})}{\sqrt{\sum_{i=1}^{n}(X_i - \bar{X})^2}\sqrt{\sum_{i=1}^{n}(Y_i - \bar{Y})^2}} \tag{1.7}$$

where X_i and Y_i: the coordinates of the ith observation, $\bar{X}$: the sample mean of the X_i values, $\bar{Y}$: the sample mean of the Y_i values, and n: the sample size.

The value of correlation coefficient r always lies between −1 and 1. Value of $r = 1$ indicates a positive linear relationship between the variables X and Y, which means that as X increases, Y increases linearly and that as X decreases, Y decreases linearly.

Percent Points Score (PPS)

Further, in the study, the status of all the issues under each component of quality practices implementation and various performances indicators, and their effects on competitive positioning of the MSME sectors Percent Points Score (PPS) for each set of questions, which reflect different issues under each component has been calculated. The PPS score reflects as to how well the area (issue) represented by that question is being looked after in the industry. This criterion is reported in earlier research studies (Nanda and Singh 2009). Percent Point Score (PPS) is calculated, according to formulae given in Table 1.4.

1.6 Literature Review on National and International Studies

Quality management has today become an all-pervasive management philosophy, which has found its way into almost every sector of business domain. After the early hype and passion for quality, it has become an established field of study with the introduction of total quality management in early 1980s. Since then, TQM has been implemented around the world in almost in service as well as manufacturing organizations. The implementation experience shared by the industry practitioners and academicians has shown a wide range of benefits of TQM. It has helped many businesses to develop competitive advantage and achieve success. TQM as a methodology had successfully drove organizations toward the continual improvement in their products and services to meet the requirements of the end customer. This holistic approach focuses not only on building an organizational culture to foster uninterrupted improvement, self-evaluation and company-wide engagement but also aims at achieving high customer satisfaction called "customer delight." Analogous to agile software development, in which success is measured by delivering a product or service, as quick as possible it aims to identify, evaluate and deliver to the customers the right products with replicating the cycle. This "collect-check-act-repeat" cycle leads to higher levels of commitment from the teams, which ultimately leads to a higher quality product or service offered to consumers.

Quality management has been developed around a number of critical factors, tools and techniques. This section provides the systematic, organized and structured detailed literature review with respect to research in the area of quality management initiatives in SMEs being analyzed by various researchers from India as well as international level, as summarized in Table 1.5. The studies summarize the key concepts related to quality management in literature in past two decades.

1.7 Concluding Remarks

In this chapter, we provided the introduction about MSMEs, their classification and importance of quality management in MSMEs. We also traced the evolution of quality management. Furthermore, we discussed about the types of statistics, methods of data collection, different types of scales and measures of central tendency. We also presented the summary of literature review where several researchers nationally and internationally have worked on identifying the factors, which contribute toward understanding and implementations of quality management practices. It has been observed that the studies are being conducted on the implementation of quality management practices, and quality standard certification in various key MSME sectors. Various research studies share the company's experience after implementing quality management practices and enlist various critical success factors, which can help MSMEs in the adoption of quality practices.

Table 1.5 Summary of literature studies

S. no	Year	Name of journal	Author details	Work done
1	2001	Total Quality Management, Vol. 12, No. 7&8, 2001, 932–938	Ana B. Escrig T., Juan C., Bou L. and Vicente R. P	Authors identified and examined the variables that affect the QM initiatives in a firm using a structural equation modeling approach
2	2001	IEEE Transactions on Engineering Management, Vol. 48, No. 4, pp. 57–62	Sanjay L. Ahire and T. Ravichandran	Relationship among the various elements of TQM is analyzed and an integrated TQM implementation approach is suggested for the effective TQM implementation in firm
3	2003	Omega, Volume 31, Issue 2, Pages 141–154	Kee-hung Lai and T.C.E Cheng	Author explored the quality initiatives of various industries and examined the links between QM implementation and quality outcomes
4	2004	International Journal of Production Economics, Volume 92, Issue 3, pp. 267–280	J. J Tari and V Sabater	Work carried is an empirical study to verify the importance of quality tools and techniques for TQM improvement and their effect upon TQM results
5	2006	TQM Magazine, Vol. 18, No. 4, pp. 410–23	Ivanovic, M.D. and Majstorovic, V.D	Authors developed a model for the assessment of quality management levels in manufacturing systems
6	2006	Journal of Fashion Marketing and Management, Vol. 10, Iss: 2, pp. 195–208	Byoungho Jin, Hwy-Chang Moon	In their work, authors used diamond approach based upon set of determinants to study the competitiveness of Korea's apparel industry

(continued)

Table 1.5 (continued)

S. no	Year	Name of journal	Author details	Work done
7	2006	Total Quality Management & Business Excellence, 17:5, 555–566	Prajogo, Daniel I. and Brown Alan	Focusing on ISO 9000 and TQM as representing different ways of introducing quality, the work in this paper investigated whether embarking on more clearly specified approaches to quality, such as ISO 9000 first or alternatively TQM, makes a difference to quality management practices and performance
8	2006	International Journal of Productivity and Performance Management, Vol. 55, No. 7, pp. 539–54	Lewis, W.G., Pun, K.F. and Lalla, T. R.M	Authors explored soft versus hard factors for TQM implementation in small and medium-sized enterprises and performed empirical investigation of the hard and soft criteria of TQM in ISO 9001 certified small- and medium-sized enterprises
9	2007	Journal of Materials Processing Technology 186, 207–213	Koc T	Impact of ISO 9000 quality management system on SMEs is analyzed and for this purpose ISO 9000 firms are compared with the non-ISO 9000 certified firms
10	2007	International Journal of Quality & Reliability Management, Vol. 24 No. 1, 2007, pp. 78–102	Jitpaiboon and Rao S. S	Authors used meta-analysis approach were used to study 421 items relating to QM practices in 50 refereed articles
11	2008	International Journal of Production Research, Vol. 46, No. 13, pp. 358–3613	Karim M.A; Smith A. J. R. and Halgamuge, S	Authors examined empirical relationship between some manufacturing practices and performance in SMEs

(continued)

Table 1.5 (continued)

S. no	Year	Name of journal	Author details	Work done
12	2008	International Journal of Productivity and Performance Management Vol. 57, No. 5, pp. 420–423	Antony J	Work presented by authors aimed to present the viewpoints from leading practitioners on the subject "Can Six Sigma be effectively deployed in small and medium-sized enterprises (SMEs)?" Results of the study clearly indicate that Six Sigma, a QM technique is equally applicable to both large corporations and small companies
13	2008	Journal of Operations Management, 26, 503–520	R.Sroufe and S Curkovic	Exploring the implications of ISO 9000:2000 and provides support for the idea that ISO 9000:2000 itself does not provide competitive advantage
14	2008	Journal of Global Competitiveness, Vol. 18, Iss: 1/2, pp. 20–28	Attila Chikan	Authors presented review on competitiveness and developed a general research model for National and firm competitiveness
15	2008	Benchmarking: An International Journal, Vol. 15, Iss: 5, pp. 525–547	Rajesh K. Singh, Suresh K. Garg, S.G. Deshmukh	Authors conducted a detailed literature review on Strategy development by SMEs for competitiveness
16	2008	Journal of Strategy and Management, Vol. 1, Iss: 1, pp. 57–71	Alan M. Rugman, Chang Hoon Oh	Authors worked upon the international competitiveness of Asian firms
17	2008	International Journal of Quality & Reliability, Vol. 25, No. 7, pp. 694–708	Fening, F., Pesakovic,G. and Amaria, P	Authors studies relationship between quality management practices and the performance of small and medium size enterprises (SMEs) in Ghana

(continued)

Table 1.5 (continued)

S. no	Year	Name of journal	Author details	Work done
18	2009	International Journal of Quality & Reliability Management, Vol. 26, No. 9, 2009, pp. 865–880	G David and Sharma B	Authors performed investigation of the hard and soft quality management factors of Australian SMEs and their association with firm performance
19	2009	International Journal of Productivity and performance management, Vol. 58, No. 3, pp. 254–273	Kumar, S and Sosnoski, M	The purpose of this study was to examine the shop floor quality issues using DMAIC approach
20	2010	Vol. 14, No. 2, pp. 54–65, Measuring Business Excellence	Seema Sharma and Milind Sharma	Authors presented state-wise cluster study for analyzing the technical and scale efficiency of small industries in India
21	2010	International Review of Business Research Papers, Vol. 6. No 6, pp. 164–173	S. Sahran M Zeinalnezhad and M. Mukhtar	Conducted an empirical study that aims to explore current implementation of management tools and advanced improvement techniques within some Malaysian SMEs in order to enhance a deeper understanding of quality management
22	2010	International Journal of Productivity and Quality Management, Vol. 5, No. 2, pp. 200–212	S.V. Lakhe and Deshmukh	Authors conducted a study aimed at identifying the Six Sigma awareness among SMEs of central India and concluded that Six Sigma has helped many industries, small and big to achieve phenomenal success
23	2010	Business Process Management Journal, Vol. 16, No. 2, pp. 226–243	Pandey VC, Garg S.K, Shankar R	It is observed that information sharing has significant impact on the competitive strengths of the manufacturer in order winning parameters like cost effectiveness and service level

(continued)

Table 1.5 (continued)

S. no	Year	Name of journal	Author details	Work done
24	2012	Journal of Operations Management, Vol 3 (4), Pages 295–315	Dong-Young Kim, V Kumar and U Kumar	Author examined the associations among different QM practices and investigate, which QM practices directly or indirectly relate to five types of innovation: radical product, radical process, incremental product, incremental process and administrative innovation
25	2013	Total Quality Management, Vol. 24, No. 1, 91–107	S. Milunovic and J. Filipovicb	Author demonstrated a modeling methodology for the QM projects in manufacturing industries in the Republic of Serbia and by conducting a survey of SME's through questionnaire
26	2013	Industrial Mgmt & Data Systems, Vol. 113, Iss: 6, pp. 856–874	Matthias T., Moacir G. Filho, Mark Stevenson, Lawrence D. F	Authors analyzed competitive priorities of small manufacturers in Brazil
27	2013	Total Quality Management, Vol. 24, No. 5, pp. 607–618	Brkic, V. K., et al.	This study empirically examined the impact of quality tools on business performance in 119 industrial firms. The relationship between prime group of quality tools, i.e., tools for reviving current conditions, for analysis current conditions and for production planning and control, were tested with regression analysis

(continued)

Table 1.5 (continued)

S. no	Year	Name of journal	Author details	Work done
28	2014	Quality Reliability and Engineering International, Vol 30, pp. 745–765	Sharma R K and Sharma R G	To investigate improvement in manufacturing performance of SMEs, authors developed an integrated model based on Six Sigma and TPM framework Various tools such as Pareto analysis, fish-bone diagrams, histograms, FMEA, control charts and process capability plots are used for analysis
29	2014	Measuring Business Excellence, Vol. 18, No. 4, pp. 86–103	Sharma, R., and Kharub, M	A conceptual framework with straightforward application of SPC to attain competitive positioning in SMEs is presented and concluded that if quality tools. It is concluded that management support and knowledge attained through proper training, the SMEs can establish their market position by enhancing the quality and productivity of their product/process
30	2014	International Journal of Production Research, 52:21, 6482–6495	Kumar M et al.	Studied the status of Quality Management practices in manufacturing SMEs by performing a comparative study between Australia and the UK
31	2016	The TQM Journal, 28(3), 338–359	Sinha, N. et al.	Studied the effect of TQM principles on performance of Indian SMEs: the case of automotive supply chain

(continued)

Table 1.5 (continued)

S. no	Year	Name of journal	Author details	Work done
32	2017	Competitiveness Review: An International Business Journal, ISSN: 1059–5422, Vol. 27 (2), pp. 132–160	Sharma, R., and Kharub, M	Authors performed comparative analyses of competitive advantage using Porter diamond model (the case of MSMEs in Himachal Pradesh)
33	2018	*J Glob Entrepr Res* **8,** 28. https://doi.org/10.1186/s40497-018-0115-5	Mukherjee, S	Authors presented challenges to Indian micro small scale and medium enterprises in the era of globalization
34	2018	International Journal of Quality & Reliability Management, Vol. 35, No. 9, 2018, pp. 1920–1940	Kharub, M., Limon, S., and Sharma, R. K	The application of quality tools in effective implementation of HACCP: an empirical study of food and pharmaceutical industries
35	2019	Quality Management and Quality Control-New Trends and Developments, Paulo Pereira and Sandra Xavier, IntechOpen. https://doi.org/10.5772/intechopen.83550	Ayon Chakraborty	Authors adopted a survey-based approach to understand the established quality management practices in Indian SMEs
36	2020	Total Quality Management & Business Excellence ISSN: 1478–3363, 2020, Vol 31, Issue-3–4, Pages: 312–341	Sharma, R., and Kharub, M	An Integrated Structural Model of QMPs, QMS and Firm's Performance for Competitive Positioning in MSMEs
37	2020	Innovative Product Design and Intelligent Manufacturing Systems. Lecture Notes in Mechanical Engineering. Springer, Singapore	Singh S.K., Mohanty A.M	Authors identified the potential barriers for successful implementation of sustainability for Indian SMEs
38	2020	Industry 4.0 for SMEs. Palgrave Macmillan, Cham	Zidek K. et al.	Authors discussed digitization of quality control operations with cloud platform computing technologies in SMEs

References

Anupam D, Himangshu P, Fredric W (2008) Developing and validating total quality management (TQM) constructs in the context of Thailand's manufacturing industry. Benchmarking Int J 15(10):51–72

Bhaumik SK, Driffield N, Zhou Y (2015) Country specific advantage, firm specific advantage and multi-nationality–source of competitive advantage in emerging market: evidence from the electronics industry in China. Int Bus Rev

Chakraborty A (10 Apr 2019) Quality management practices in Indian SMEs, Quality Management and Quality Control-New Trends and Developments, Paulo Pereira and Sandra Xavier, Intech Open. https://doi.org/10.5772/intechopen.83550

Claver E, Tari JJ, Molina JF (2003) Critical factors and result of quality management: an empirical study. Total Qual Manag 14(1):91–118

Ministry of MSME's (2017) Annual report 2017/18. New Delhi: Government of India. https://msme.gov.in/sites/default/files/MSME%20ANNUAL%20REPORT%202016-17%20ENGLISH.pdf

Nanda T, Singh TP (2009) An assessment of the technology innovation initiatives in the Indian small manufacturing industry. Int J Technol Policy Manag 9(2):173–207

Juran JM (1988) On planning for quality. Collier Macmillan, London. New York: Free Press

Khanna HK, Sharma DD, Laroiya SC (2011) Identifying and ranking critical success factors for implementing of total quality management in the Indian manufacturing industry using TOPSIS. Asian J Qual 12(1):124–138

Kumar V, Choisne F, de Grosfoir D, Kumar U (2009) Impact of TQM on company's performance. Int J Qual Reliab Manag 26(1):23–37

Mukherjee S (2018) Challenges to Indian micro small scale and medium enterprises in the era of globalization. J Glob Entrepr Res 8:28. https://doi.org/10.1186/s40497-018-0115-5

RBI (2017) Handbook of statistics on Indian economy. Mumbai: Reserve Bank of India. https://rbidocs.rbi.org.in/rdocs/Publications/PDFs/0HANDBOOK2017C9CF31D4B78241C9843272E441CD7010.PDF

Porter M (1990) The competitive advantage of nations. Simon and Schuster. Free Press, New York

Sharma M, Kodali R (2008) TQM implementation elements for manufacturing excellence. TQM Mag 47(8):1281–1299

Talib F, Rahman Z, Qureshi MN (2013) An empirical investigation of relationship between total quality management practices and quality performance in Indian service companies. Int J Qual Reliab Manag 30(3):280–318

Chapter 2
Quality Management Tools and Standards

Highlights

- *Quality management basically comprises of three main components: quality control, quality assurance and quality improvement. To take account of these three, there are various tools cum techniques, which are enormously used to collect, measure and analyze the data related to product's manufactured and processes being adopted by the organizations.*
- *The tools like Check sheet, 5 S and housekeeping, Kaizen, Scatter diagram, Histogram, Pareto Diagram are easy to use and can be adopted for getting effective results whereas other tools such as FMEA, Six Sigma, Kanban, QFD, CRM need extensive training and expertise, with proper management support/commitment for effective implementation.*
- *The available quality management standards are usually generic in character applicable for all organizations or for all organizations within a particular industry type like automobile, telecommunications,* etc.
- *This chapter provides details to*
  - *Various quality management tools along with their advantages and necessary procedures to use them.*
  - *Quality System Standards and implementation plan.*

2.1 Introduction

Quality management basically consists of three main components: quality control, quality assurance and quality improvement. To take account of these three, there are various tools cum techniques that are enormously used to collect, measure and analyze the data related to product's manufactured and processes being adopted by the organizations. Adopting quality management practices means call for cultural

R .K. Sharma, *Quality Management Practices in MSME Sectors*,
Springer Tracts in Mechanical Engineering,
https://doi.org/10.1007/978-981-15-9512-7_2

transformation that requires improvement at all levels of organization. This chapter discusses various tools and standards for successful implementation of quality management programs in industry.

2.2 Quality Management Tools

Quality management tools and techniques impose a certain set of disciplines in which they assist organizations to improve progressively by building and sustaining positive interactions among the processes or products. Examples of some commonly and widely used tools are McQuarter et al. (1995), Dale and McQuater (1998), Brkic et al. (2013):

- Cause and effect diagram;
- Relationship diagrams;
- Pareto analysis;
- Histograms;
- Flowcharts;
- 5s;
- Scatter diagram;
- Process flow diagram;
- Control chart;
- Check sheet.

A quality technique, whereas, has a broader application than a single tool and it can be understood as a set of individual tools. It required more thoughts, skills and training to implement a technique correctly. Examples for techniques are Goetsch and Davis (2003), Tari and Sabater (2004), Talib et al. (2013); Anderson et al. (2007); Juran (1974); Garvin (1984); Chambers et al. (1983); Juran and Gryna (1993); Crosby (1979); Maritan (2014); Cristiano et al. (2001); Harland (1996); Gartner: Market share (2018) respectively.:

- Statistical process control;
- Benchmarking;
- Quality function deployment;
- Failure mode and effect analysis;
- Design of experiments;
- Six Sigma;
- Supply chain management (SCM);
- Poka-Yoke;
- Kanban;
- Kaizen;
- Enterprises resource planning.

The upcoming sections shall present in details about the tools and techniques in the domain of quality management, procedural steps and main benefits of using them.

2.2.1 *Cause and Effect/Fishbone/Ishikawa Diagram*

It is the method that is used to investigate and demonstrate the view about the causes, which results in process variation. The method is known in literature by different names such as Ishikawa diagram, Fishbone diagram and Root Cause Analysis. The name Ishikawa diagram is after the inventor, Ishikawa (1976). The main motive of application of the tool is to disembark the various sources/causes that contribute most extensively toward the problem under examination. These causes are then taken for initiating improvement actions. The method also portrays the interactions among the wide range of potential causes of the effect. Figure 2.1 presents cause and effect analysis. The diagram shows six possible causes, i.e., cause 1, cause 2, cause 3, cause 4, cause 5, cause 6 to a particular event or problem on hand.

These causes are connected to a horizontal line which on right-hand side connects to a problem. After identification of main causes, the subcauses are added to the bones on the main bone represented by the horizontal line. The fundamental notion in the cause and effect analysis is that the "basic problem" of concern is entered at the right side of the diagram at the end. The major possible causes of the problem (known as the effects) are listed on the horizontal line. For instance, for a typical process the "4-M" is used as a reference point: "Materials," "Machines," "Manpower" and "Methods" to find out the deviation, which may influence the product/process performance. In the quality circles, the members usually brainstorm the problem on hand and use this tool to add all potential causes to the main "bones" (called the 4Ms) and more specific causes to the "bones" on the main "bones." For better understanding of the problem, this subdivision continues till the problem is figured out completely. When the diagram is completed, one has complete picture of all the potential root causes for the designated problem. The diagram can be used by individuals or teams; most

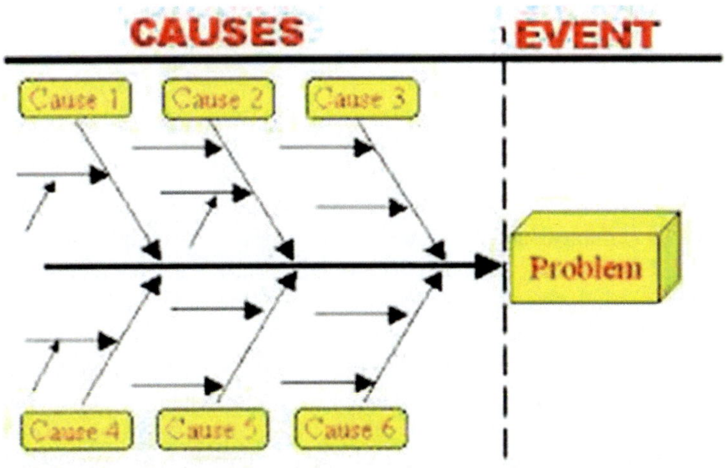

Fig. 2.1 Cause and effect diagram

likely in quality circle groups. The three key functions of cause and effect diagrams are (i) finding cause details, (ii) analysis of deviation and (iii) analysis of process. The cause details are mainly obtained through a brainstorming sessions in which all probable causes are pointed out to show their influence on the problems (or effect) on hand. In deviation analysis, the reasons for deviations of each major cause are thoroughly examined by listing out subcauses and studying their likely effect on the quality characteristics, so called the event or problem in question. This brings the reasons for variability with in a process/system/subsystem. When such diagrams are built for process analysis, the stress is placed on recording various resulting causes in the succession in which the different procedures are carried out. The various steps for constructing cause and effect diagram are discussed as under in Sect. 2.2.1.1, respectively.

2.2.1.1 Construction Steps

I. Define the problem or the effect that needs to be studied or examined.
II. Formulate the team of individuals to carry out the analysis. Often, the teams will discover the probable cause through brainstorming.
III. Draw the effect box on right-hand side and the center line.
IV. Identify the probable cause categories and join them as boxes connected to the centre line.
V. Identify the probable cause and classify them into the categories and create new subcategories, if necessary.
VI. Ranks order these possible causes in order to recognize those that seem almost certainly to affect the problem.
VII. Initiate the remedial actions.

Figure 2.2 shows the cause and effect analysis diagram that is used to separate out various quality cost elements under each cost category.

2.2.2 Check Sheet

Check sheet is a simple data collection form that consists of multiple categories with definitions to enter the data related to the problem. Data are entered on the form with a simple tally mark that represents the occurrence of the activity. The main purpose of the tool is to facilitate the data collection process.

A distinctive feature of a check sheet is that data are recorded on it by making checks on it after dividing the check sheet into various regions, and recoding marks based on observations in those divided regions. Finally, the data are read by observing the region and number of checks/marks in that region. This tool can be used for various

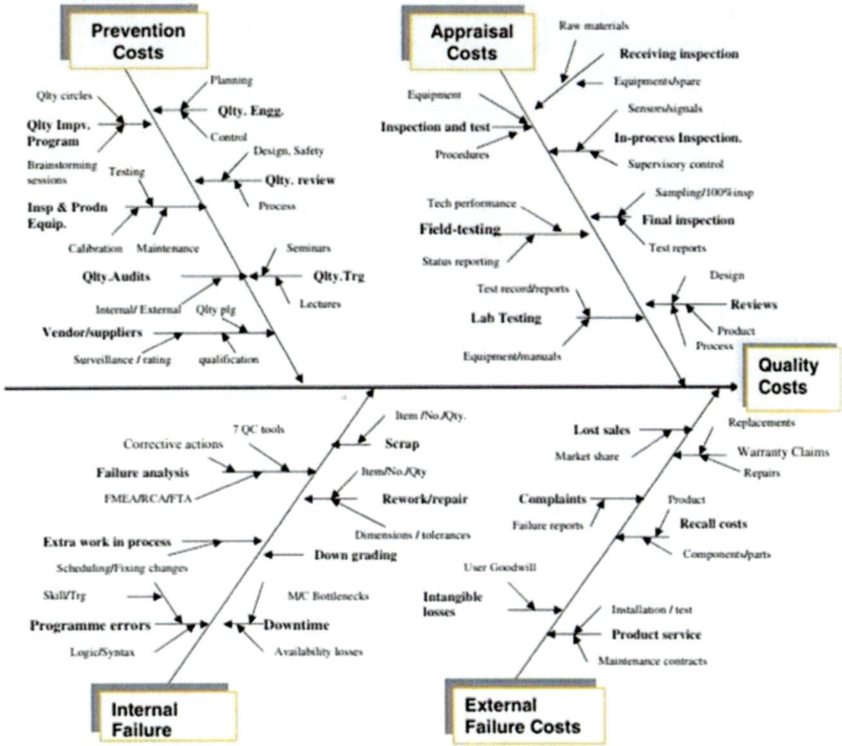

Fig. 2.2 Cause and effect diagram for quality costs (Sharma et al. 2007)

activities. For instance, for checking the likelihood of different types of defects in castings or to count the expected items, e.g., the times the telephone interruptions observed in a week, a sample checklist is shown in Table 2.1. The various reasons for interruptions are listed in the table and a tally mark is used to check beside each reason when it appears.

The four different ways in which check sheets are classified as under:

Table 2.1 Sample check sheet

Reason	Day					
	Mon	Tues	Wed	Thur	Fri	Total
Wrong number	IIIII	II	I	IIIII	IIIII II	20
Info request	II	II	II	II	IIII	10
Call from manager	IIII	II	IIIII II	I	II	18
Total	11	6	10	8	13	48

A. **Classification**: It is used when characteristics or traits such as a defect or modes of failure are required to be classified into different categories.
B. **Measurement Scale**: It is divided into intervals, and measurements are shown by the examination of a proper interval.
C. **Location**: It is used when we have to indicate the physical location of a trait on a picture of a part or item.
D. **Frequency**: It is used when we have to indicate the appearance or nonappearance of an attribute or different attributes.

2.2.2.1 Cheek Sheet Procedure

1. Define the problem or decide which event is to be observed. Develop operational definitions for the problem under investigation. Decide about the data needs, its collection and time duration for which it needs to be collected.
2. Design the form or template with various sections. Set it up so that data can be recorded simply by making tally marks or similar symbols for analysis.
3. Label all the spaces appropriately on the form.
4. Every time when the targeted event or problem occurs, make record of data using tally marks or similar symbols.
5. Perform pilot test of the check sheet for a shorter period to ensure that it is used to record the data in suitable manner and is simple to use for further analysis.

2.2.2.2 Advantages of Using Check Sheet

(i) The check sheet is a straightforward and effective way to demonstrate data.
(ii) It provides means to understand the problem in simple manner as it provides a uniform way to collect the data.
(iii) In lean six sigma, it is widely used to distinguish the judgment from facts in the define and measure phase of the DMAIC process.

2.2.3 Control Charts

In the production process, the managers have to examine whether their products/processes meet the desired specifications or not. In the broadest terms, the two biggest "enemies" of product quality are (i) dispersion from the mean or target and (ii) deviation around targeted mean. If the variation occurs *outside* the control limits, then it is because of special causes. But, if the variation occurs *within* the control limits, then it is because of common causes. Common causes are inbuilt to the system and variation because of these causes is not eliminated totally. So, a production process operating under a stable system of common causes is said to be under statistical control. Deming and Juran believed that about 85% of all problems are because of common causes and thus can be overcome by management initiatives. The variations due to special causes can be controlled. For example, the operator error may be because of inadequate training or use of improper technique, which can be corrected by proper training or using appropriate techniques. Other example of special cause is inferior quality shipments from vendors. According to Deming, 15% of all problems are because of special causes. Initiatives by management and employees would reduce the likelihood of special causes. Various methods/charts under quality control can be used to monitor online or in-process quality of an on-going production process. Managers need to be accustomed to elementary concepts such as sampling distribution and the basics of the normal distribution. Section 2.2.3.2 describes the procedure for constructing the control charts.

Example: Let the plant manager wants to control the size of piston rings, with the assumption that the mean and variance of the process do not change and the succeeding sample means follow normal distribution around the targeted mean. Moreover, without going through the derivation, we know from the central limit theorem that the distribution of sample means will possess a standard deviation of sigma over the square root of n (the sample size). It follows that about 95% of the sample means will lie within $\mu \pm 1.96 * \sigma/\sqrt{n}$ the limits (for details refer to Montgomery, Douglas, C. Introduction to Statistical Quality Control, Sixth Edition). In general, it is common to replace ± 1.96 with ± 3 sigma limits to accommodate approximately 99% of the observations, respectively.

2.2.3.1 Common Types of Control Charts

The types of control charts are categorized based upon the quality characteristics that they are going to examine, i.e., charts for variables and charts for attributes. The variable charts are usually used for continuous measurements, such as diameter, height, length, voltage, etc. The attribute charts are used to measure discrete data, in the form of counts for instance, the number of defects or number of customer complaints. The attribute charts are usually used in service organizations to sustain and improve quality improvement efforts as the quality parameters in such organizations are not ascertained on a numerical scale.

In particular, the subsequent charts are commonly used for controlling variables:

- **X-bar chart**. The chart is used to graph the sample means with an aim to exercise control over the target value of a variable (e.g., piston ring dimensions, material strength, shaft diameter, etc.).
- **R chart**. The sample ranges are plotted using R chart to control the variability within the data.
- **S chart**. The sample standard deviations are obtained through S chart in order to control the variability within the data.

In order to control the quality characteristics representing the attributes of the product, the various types of charts commonly used are:

- **C chart**. The chart gives the total number of nonconformities in constant size sample. Using C chart, the number of defectives (per batch, per day, per machine, etc.) is plotted. The probability of the occurrence of nonconformities follows Poisson distribution with mean and variance equal.

The average number of nonconformities is denoted by $\bar{C}$ and is obtained from the sample observations with centerline and control limits given as under:

$$CL = \bar{C} \; UCL = \bar{C} + 3\sqrt{C} LCL = \bar{C} - 3\sqrt{C}$$

- **U chart**. Unlike C chart, it does not require a constant sample size. It is used for monitoring the situations in which the sample size varies. It plots the rate of defectives, (e.g., feet of pipe, number of batches) and can be used, for instance, when the samples are of different sizes. The number of nonconformities per unit for the *i*th sample is given by

$$u_i = \frac{c_i}{n_i}$$

where c_i represents the number of defectives in the *i*th sample, and n represents the size of the *i*th sample.

The average number of defectives per unit ($\bar{u}$) and control limits are given as

$$\bar{u} = \frac{\sum_{i=1}^{g} C_i}{\sum_{i=1}^{n} n_i}$$

$$UCL = \bar{U} + 3\sqrt{\frac{\bar{U}}{n_i}}$$

$$LCL = \bar{U} - 3\sqrt{\frac{\bar{U}}{n_i}}$$

- **P chart**. The chart gives the proportion of nonconforming items or percentage of defectives (per batch, per day, per machine, etc.). The proportion nonconforming for a given sample , $\bar{p}$ is stated as:

$$\bar{p} = \frac{x}{n}$$

where x: represents nonconforming items in the sample and n represents the sample size.

Further, based upon 3-sigma limits, control limits for P-chart are given as:

$$UCL = \bar{p} + 3\sqrt{\frac{\bar{p} + 3(1 - \bar{p})}{n}}$$

$$LCL = \bar{p} - 3\sqrt{\frac{\bar{p} + 3(1 - \bar{p})}{n}}$$

where $\bar{p}$: represents is proportion nonconforming, and n: represents sample size.

The control limits in P chart are based on the binomial distribution in which the likelihood of getting a nonconforming item remains constant. Moreover, a p chart can be applied to investigate the quality of an operator or machine, a work center, or an entire plant. Thus, this chart is mainly appropriate for analyzing those situations where the likelihood of defectives is more frequent (e.g., we anticipate to get a nonconforming items so called defects in a foundry to be greater than 4% of the total units produced).

- **Np chart**. In np chart, we normally count the number of defectives or nonconforming items in the sample. It is assumed that these nonconforming items follow a binomial distribution. If the sample size changes, the centerline and control limits change as well. Thus, an np chart is not recommended for the situation where the sample size varies because making inferences in such situations becomes difficult.

  The control limits for *np* chart are given as

$$\text{CL} = n\,\bar{p},$$

$$UCL = n\,\bar{p} + 3\sqrt{n\,\bar{p}\,(1 - \bar{\text{p}})}, \text{ and}$$

$$LCL = 3\sqrt{n\,\bar{p}\,(1 - \bar{\text{p}})}$$

where $\bar{p}$: represents the proportion nonconforming, and n: is sample size.

If the LCL has negative value, than it is changed to zero.

2.2.3.2 Guidelines to Draw Control Chart

The control charts are used to determine whether the quality improvement initiatives put a stop to particular problems or fundamental changes to the process are required. The general guidelines that may prove beneficial to implement control charts are.

1. Select the suitable control chart for your data (variable or attribute data).
2. Decide upon the suitable time period for collection of the data and to plot the results for making inferences.
3. Check the appearance of "out-of-control signals" on the chart. When one is identified, mark it on the chart and investigate the cause for further investigations. Table 2.2 provides the description to various rules for out of control situations. Figure 2.3 shows different out of control situations as per the rules described in Table 2.2.

2.2.4 *Process Flow Chart*

Flowchart provides a means of diagrammatic representation of the flow of data/raw material through information processing systems/shop floor, the sequence of operations being executed within the system/shop. It helps to identify blockages, unnecessary steps and nonvalue-added activities.

For instance, the program flowchart is similar to the blueprint of a building. As we know that an architect draws a blueprint/plan of the building before a builder initiates the construction work. On similar lines, in agile software development, a

Table 2.2 Rules for Out-of-control signals/patterns

Rule number	Rule description	Explanation
Rule 1	If only one point lies outside the control limits	In Fig. 2.3, point 16 lies above upper control limit
Rule 2	Two out of three consecutive points are on the similar side of the centerline and lies away than 2σ from it	In Fig. 2.3, point 4 represents this case
Rule 3	Four out of five consecutive points are on the similar side of the centerline and lies away than 1σ from it	In Fig. 2.3, point 11 represents this case
Rule 4	A run of eight points in a row are on the identical side of the centerline	In Fig. 2.3, point 21 represents this case
Rule 5	Noticeable steady or constant patterns that imply somewhat strange about the process	–
Rule 6	Keep on plotting data, for every new data point which is plotted, test for latest out-of-control indication	–

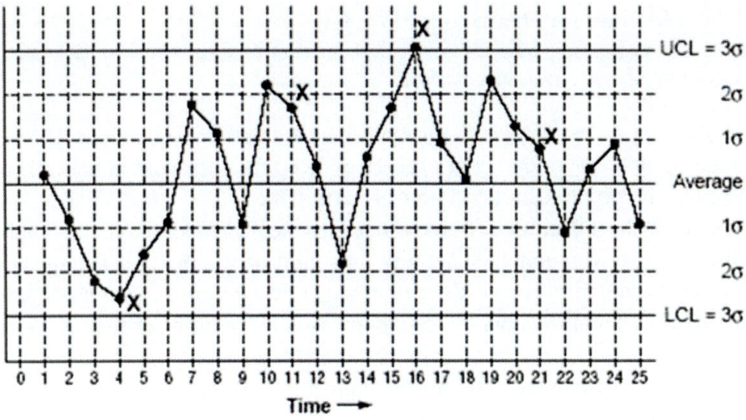

Fig. 2.3 Out of control signals (*Source* www.asq.org)

software developer draws a flowchart based upon customer requirements prior to development or release of a computer program. The flowchart is made by taking into consideration the set rules and using typical symbols approved by the American National Standard Institute, Inc. The five process chart symbols with description are shown in Table 2.3, respectively.

Made in the initial stages of work, flowcharts assist in sharing information among the programmers and business people/management. The flowcharts demonstrate the length of process from the beginning of an activity to the end of the activity. In software developing companies, once the flowchart is ready, it becomes easy for the developers to make the program in any high level language. A thorough description of the flowchart is called the process map, which recognizes process inputs in form of material, equipment, personnel, and method and process outcomes in terms of product or service.

Table 2.3 Process chart symbols

Symbol	Letter	Description
O	O [Operation]	It refers to changes in the physical characteristics or chemical composition of the material
IL	I [Inspection]	It is the activity of checking the quality or the quantity of the product
→	M [Transport]	To transport the material from one position to another
D	D [Delay]	It refers to delay in the process when material cannot move to the next activity
∇	S [Storage]	It refers to storage when the material is kept in stores

2.2.4.1 Guidelines for Drawing a Flowchart

The general guidelines for drawing flowcharts are as:

1. Identify the process for which the flowchart is required to be made.
2. Decide upon the boundaries of process by keeping into account the information related to the process start and process termination.
3. Brainstorm the different activities that have to be undertaken for completion of the process. Mention each activity on a card or sticky note. Though order of operations is not important at this point, but brainstorming the order of operations may help the team members to retain information with respect to all necessary steps.
4. In this step, organize the activities in appropriate order.
5. When all activities are incorporated and everyone in team agrees about the proper order of activities, draw arrows to demonstrate the process flow.

2.2.4.2 Advantages of Using Flowcharts

Following are the benefits of developing the flowcharts:

1. Communication: They provide improved way of communication of the logic of a system to all the stakeholders in a process.
2. Proper documentation: They provide better documentation of the plan, which is further required for completing different processes.
3. Efficient Coding: They act as a guide or draft during product design and development phase.
4. Effective analysis: By using the flowchart, the problem can be investigated more efficiently.
5. Accurate Debugging: The flowchart assists in debugging process. The maintenance of operating program becomes easier by using flowchart.

2.2.5 Histogram

It provides a means to *display* the data graphically (Fig. 2.4) in such a manner that the feature under investigation is categorized into classes. In a frequency histogram, the vertical axes signify the number of observations. It presents the percentage of cases that fall into each of several categories and allows us to draw conclusions about the state of the process. It differs from a conventional bar chart in which the height of bars is used to represent the dimensions but in case of histograms it is the area of the bar that denotes the value, not the height, a crucial difference when the groups are not of consistent width. The groups (bars) are generally specified as nonoverlapping intervals of some variables. Figure 2.4 shows the histogram of quality defects per hour in a foundry with their frequency on vertical axis.

Fig. 2.4 Histogram

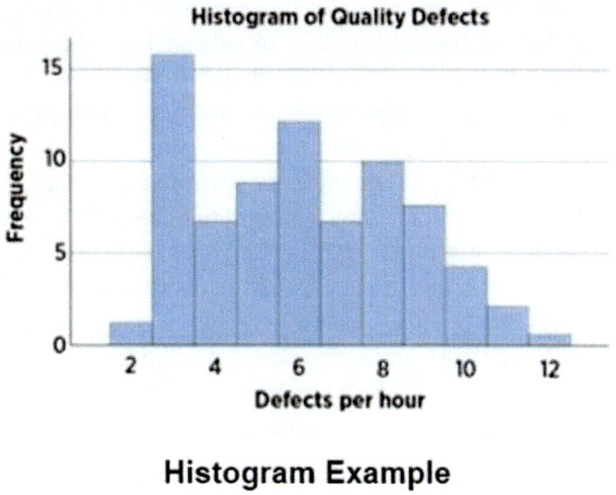

2.2.5.1 Steps to Construct a Histogram

Following steps are used to construct histograms

Step 1: At first divide the range of the data into identical size of classes. Depict and make your *x* and *y* axis. For the example above, the *x*-axis would be named as "defects per hour" and the *y*-axis would be named as "relative frequency %."

Step 2: Show the class intervals along the *X*-axis on an identical scale.

Step 3: Show the frequencies along the *Y*-axis on an identical scale.

Step 4: Make rectangles with class intervals as bases and subsequent frequencies as heights.

2.2.5.2 Advantages of Histogram

1. It displays large amount of data and helps track the trend.
2. It can be used to show the relative frequency of occurrence of the data, thus acts as powerful tool for communicating information.
3. They show the upcoming performance of a process.

2.2.6 Pareto Diagrams

Named after Vilfredo Pareto, an Italian economist (1848–1923), the use of Pareto diagram in quality assurance was popularized by Juran and Ishikawa, quality gurus. It is a particular kind of bar chart in which the values being plotted are prearranged in decreasing fashion.

The chart provides variables required for setting priorities. It helps to arrange and display the information to demonstrate the relative significance of causes of problems. It is a vertical bar chart that organizes the items from the maximum to the minimum, relative to some quantifiable metrics, i.e., occurrence, cost, time. The diagram is based on the Pareto principle, which states that the greater part of wealth in the world, is detained disproportionately by small section of the population. It supports the 80/20 rule, which defines that 80% of nonconformities or defects in production are due to 20% of the causes. Placing the items in decreasing order of occurrence makes it easy to distinguish those areas that are of utmost significance or segregating those causes that result in most of the variation. Thus, a Pareto chart helps teams to prioritize problems and center their efforts in an environment of limited resources. Figure 2.5 shows Pareto diagram for the defects in a shirt. The diagram shows following salient features:

- In the example presented in Fig. 2.5, each bar shows a type of defect. The bars are shown in decreasing pattern (from highest to lowest). The height of the bar shows the unit of measure say, the rate of occurrence. The three topmost defects, i.e., button defects, pocket defects and collar defects, contribute toward 83% of the defects.
- When the cumulative percentage line (shown with red color) starts to flatten, the resulting defect types do not earn much focus, because addressing them will not be so fruitful because of their less frequency or percentage.

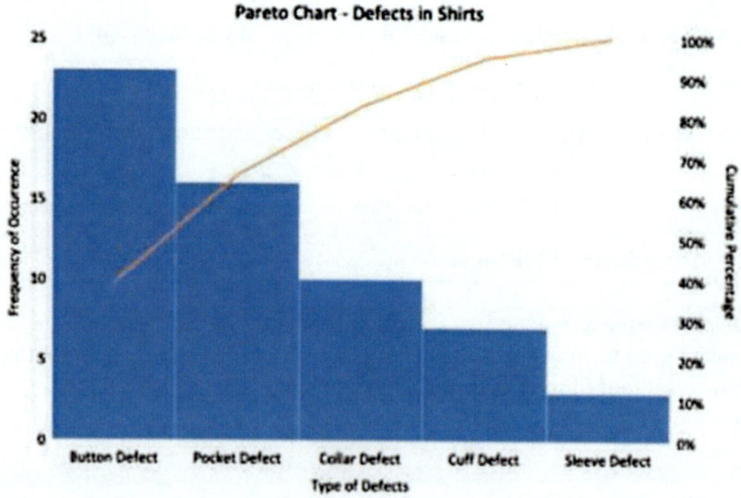

Fig. 2.5 Pareto diagram

2.2.6.1 Guidelines to Construct Pareto Diagram

Step 1: Prepare a list of problems, items, or causes for which the comparisons are to be made.

Step 2: Develop a standard measure for the comparisons of the items on the basis of:

- Their rate of occurrence (e.g., utilization, complications, errors)
- Their longevity, i.e., how long it takes? i.e., the time
- The number of resources used by them, i.e., the costs involved.

Step 3: Select the period of reference, i.e., the time for collection of the data related to the problem.

Step 4: Use tally marks, for each objects/defects/causes, how frequently they occurred? Then sum up them to find out the average total for all objects/defects/causes. Find the percentage of each item in the average total by taking the sum of the objects, dividing it by the average, and finally multiplying by 100.

Step 5: Prepare the list of the items/defects/causes being compared in descending order from most common to the least common.

Step 6 Items ranging from highest to lowest frequency are placed on the horizontal axis and the left side of vertical axis is shown with the frequency and the right vertical axis is shown with the cumulative percentages.

Step 7: Develop a line graph of the collective percentage of the total.

Step 8: Evaluate the graph by categorizing the items with more contribution toward the problem.

2.2.6.2 Advantages of Pareto Diagram

I. The Pareto diagram can be used to discover the major problems and also the minor ones.
II. It depicts the importance order of the inputs in a simple, visual format.
III. It helps in setting the agenda for corrective actions.
IV. It helps the team members to focus on the inputs with maximum impact.

2.2.7 Scatter Plot

It is used to display values for two variables for a set of data. By demonstrating the variables on each axis, one can perceive the relationship or correlation among the variables of interest. Scatter plot is also known as point graph, X–Y Plot, scatter chart or scatter gram. They are used to determine various kinds of associations among variables with a certain confidence level. These correlations may be positive, negative or null. If one variable (dependent variable) increases with increase in the other variable (independent variable), than there is said to be a positive correlation. Increase in height with increase in the size of clothes is an example of a positive correlation. Another example is an increase in the age of a child with increase in his/her weight (as shown in Fig. 2.6a. To study the correlation between the variables a line of best fit called "trend line" can be made and the equation for the correlation among the variables can be ascertained by establishing best-fit measures. For a linear association, the best-fit procedure is called linear regression. Figure 2.6b shows negative correlation between variable X and variable Y, which occurs when decrease in X results in increase in Y. For instance, a student who has many absents has a decrease in grades. Figure 2.6c presents no correlation between variable X and variable Y, which occurs when change in X has no impact on Y. For example, the length of time a person has spent on watching television has no impact on electricity bill. Thus, we

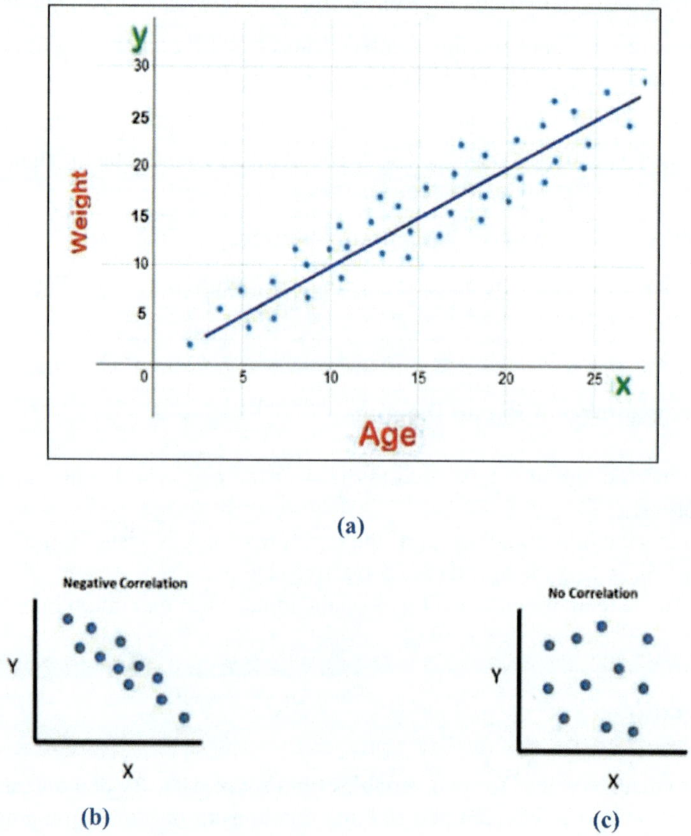

Fig. 2.6 **a–c** Scatter plots

can say that scatter plots are not one of the most often used visualization kind of charts, but they play significant role by handling large quantities of data.

2.2.7.1 Steps to Construct Scatter Plot

I. Data are collected as order pairs (x, y).
II. The horizontal and vertical scales are constructed with the higher value on the right for the x-axis and in the top for the y-axis.
III. After the scale is labeled, the data are plotted.

2.2.7.2 Advantages of Scatter Plot

The advantages of a scatter diagram are:

I. It presents the relationship and a trend in the data between variables X and Y.
II. It is the best way to show a nonlinear pattern.
III. The data ranges, i.e., highest and smallest values, can be determined.
IV. Plotting the diagram is easy. Observation and reading are straightforward.
V. It presents all data points, including least and greatest values with outliers.

2.2.8 Customer Relationship Management (CRM)

Customer relationship management is an approach to manage company's relations with existing and potential customers. It uses data analytics about customers' history with a company to improve business relations with customers, particularly with focus on customer retention and promotion of sales growth potential. Even today, this kind of an association exists between customers and retailers, craftsmen, artisans—basically in markets that are conventional and small in size. Technology for CRM includes equipment, software and communication links that organizations use to enable or improve their processes electronic point of sale (EPOS), sales force automation, customer service helpdesk, call centers, etc. Gartner in their report predicted that by 2021, CRM will be the single largest revenue area of spending in enterprise software.The four major vendors of CRM systems are Salesforce, Microsoft, SAP and Oracle.

The three major components of CRM system are:

(i) CRM Software—It consists of relational database for storage of persistent information and key software applications for handling business logistics.
(ii) Client Hardware—It could be a personal computer or handheld device for the assessment of enterprise information.
(iii) Mobile Middleware—It provides assistance in the information exchanges amidst CRM software and access devices or PCs and helps to access information across the organizational lines and locations.

2.2.8.1 CRM Software

It combines customer information and documents into a database so that business users can access and manage it easily. Other key functions of this software consist of making records of customer communications over phone calls, emails, social media or other channels on automation of different processes, and giving CRM managers the facility to measure the performance. The various automation processes under CRM are:

(i) **Marketing automation**: It helps to automate repetitive tasks with an aim to increase marketing efforts at different locations.
(ii) **Sales force automation**: It helps to track the customer communications and automate business activities essential to follow leads and draw new customers.
(iii) **Contact center automation**: It might consist of prerecorded audios that assist in customer problem-solving and information dissemination.
(iv) **Location-based services**: The location-based technology is used as a networking tool to unearth sale prospects based on geographic conditions with GPS apps.
(v) **Workflow automation**: It helps the businesses to optimize processes by streamlining routine workloads, which enables the employees to focus on innovative high-level tasks.
(vi) **Human resource management**: CRM systems help to track information related to employees, so that the internal workforce is managed more effectively.
(vii) **Analytics and Artificial intelligence (AI)**: To achieve better customer satisfaction index by analyzing user data analytics play important role. In present times, artificial intelligence technologies, such as Salesforce, have been integrated into CRM platforms to automate recurring tasks, to discover customer buying behavior and to predict future customer behaviors.

2.2.8.2 Advantages of CRM

1. Introducing CRM in organizations has shown promising results with direct advances in the bottom line, for instance, increase in sales by +30%, sales productivity by +35%, customer satisfaction by +35%, faster decision-making by + 38% and revenue by +25% (Source: *sales force relationship survey* conducted 2014–2016 among 10,500+ customers randomly selected and response sizes per question vary).
2. A CRM system identifies and adds new leads easily and quickly and classifies them with more accuracy.
3. It also helps to raise recommendations from present customers, by being considerate toward your customers shall result in cross-selling and up selling of products.

4. CRM captures information from the business entities and beyond, which is helpful to the companies to initiate improvements in contributions, finding the problems early, and their solutions.

2.2.9 Enterprises Resource Planning (ERP)

In order to be successful in increasingly competitive environment, the companies need to adopt such systems, which allow them to easily automate their business operations. An enterprise management information system integrates all facets of business enterprise from planning, purchasing, inventory, finance, sales and marketing, etc. into one comprehensive information system. The term "ERP" or "Enterprise Resource Planning" was coined by industry analyst, The Gartner Group, in 1990s. It is evolved from material requirement planning systems, which were created back in the 1960s when manufacturing-based companies used these systems to aid production line managers for improving efficiency. Today, ERP systems are integrated into all facet functions of an enterprise, with the main concern to provide managers better decision-making capabilities regarding their operations. With more popularity of ERP across the businesses, large software applications are available to aid enterprises in implementation of ERP-based systems.

A typical ERP system comprises of applications and tools that help all business areas to communicate with each other more effectively. The right kind of ERP system can help the companies to gather and record data into one central location from various key areas such as:

- Finance and accounting services.
- Human resource management.
- Customer relationship management.
- Production and operations.
- Maintenance planning and spare parts management.
- Business intelligence.
- Inventory and logistics-based systems.
- Supply Chain and E-commerce.
- Point-of-Sale (POS).

2.2.9.1 Advantages of ERP

1. It enables managers in identification of challenges, discover future prospects, and make faster decisions that affect various business areas.
2. It provides managers with instantaneous information related to operations and real-time access to their supply chain networks.
3. It provides data security, which helps the businesses to continue to function in agreement with global regulations and guidelines.

4. It helps the managers to work in partnership with each other and with third-party vendors as the information gets accessible from any location by using mobile devices.
5. ERP software makes use of reporting and forecasting tools to undertake informed decisions about the future business prospects. Hence, it helps to increase productivity.

2.2.10 Kaizen

The term "Kaizen" came from two Japanese words: Kai (improvement) and Zen (good), which means "continuous improvement." It has its origins in post-World War II, which has seen the formation of Japanese quality circles. The circles were mainly focused on preventing defects at Toyota production and were formed in reaction to American management and productivity consultants, especially Deming, who emphasized that quality is controlled by the workers. In his book "Out of the Crisis," Dr Deming shared his philosophy of continuous improvement with 14 points called Deming's fourteen points: It was brought and popularized to the West by Masaaki Imai through his book *Kaizen: The Key to Japan's Competitive Success* in 1986.

In Western society, kaizen is broken down into four stages: assessment, planning, implementation and evaluation. Toyota is debatably the most famous for Kaizen application, but other companies such as Lockheed Martin, Ford Motor, Nestle etc. have used the approach successfully.

Kaizen implementation

The technique used to implement kaizen is known as the Plan—Do—Check—Act widely known as Demings PDCA cycle. In this technique, in the first phase, we plan the activity as KAIZEN implementation activity by selecting the area or the process, which needs improvement. The current state of the process is measured and all the observations related to the process are documented. The main aim of observing the current state of the process is to find out the existing wastes in the process. In the second phase called "DO," the important knowledge pertaining to the process is made available to the teams by imparting training in order to fill the gaps among team members. The teamwork in collaboration and documents the improvement plan and which is then implemented in the process. The third phase is called "CHECK" in which the adherence of planned activities for potential improvement is investigated. The final phase of PDCA cycle is the ACT in which the observed gaps found during audits are worked upon. Figure 2.7 shows the standard steps discussed above in order to implement Kaizen in the workplace.

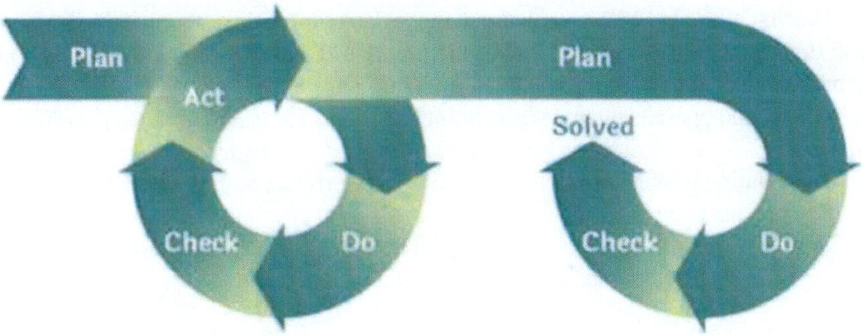

Fig. 2.7 Steps to implement Kaizen

2.2.10.1 Advantages of Kaizen

1. With its focus on steady improvement, it creates a gentle way toward change in comparison to enormous efforts.
2. It promotes close inspection of processes, which helps to reduce mistakes and wastages.
3. It also helps to reduce inspection needs because errors are reduced.
4. It helps to increase collaboration among employees, as they think ahead of the specific issues of their section.
5. Additionally, organizations with more engaged employees can achieve higher competitiveness, enhance customer satisfaction, and have a culture of solving problems through teamwork.

2.2.11 FMEA (Failure Mode and Effects Analysis)

Failure mode and effect analysis (FMEA) is a structured, bottom-up approach that begins with known possible failure modes at one subsystem level and examines the effect on the next subsystem level. Undoubtedly, all mechanical systems are composed of several subsystems, which can be further decomposed to a component level. FMEA as a recognized design approach was developed at Grumman Aircraft Corporation in the 1950 and 60s and was applied to naval aircraft flight control systems. Since then, it has been widely used as influential tool for safety and reliability analysis of products and processes in a wide range of industries particularly, aerospace, nuclear and automotive industries. In 1977, FMEA was implemented and promoted by Ford Motor Company. It leads to use of FMEA in automobile sector to evaluate and prioritize probable design and process-related failures. The main purpose of FMEA is to determine and correct the potential failure problems related to design and production.

The first phase in FMEA is related with the detection of the possible failure modes and their effects. It includes defining the possible failures of product's component, subassemblies, final assembly and its production processes.

The second phase is related with getting the scores for *likelihood of occurrence of failure* (S_f), *Severity* (*S*) *of failure, and likelihood of the nondetection of failure* (S_D) and, calculating Risk Priority Number (RPN) as a product of three probabilities, i.e., S_f, *S*, S_d respectively.

Figure 2.8 presents the general procedure for conducting FMEA study. In short, the steps are as discussed as under.

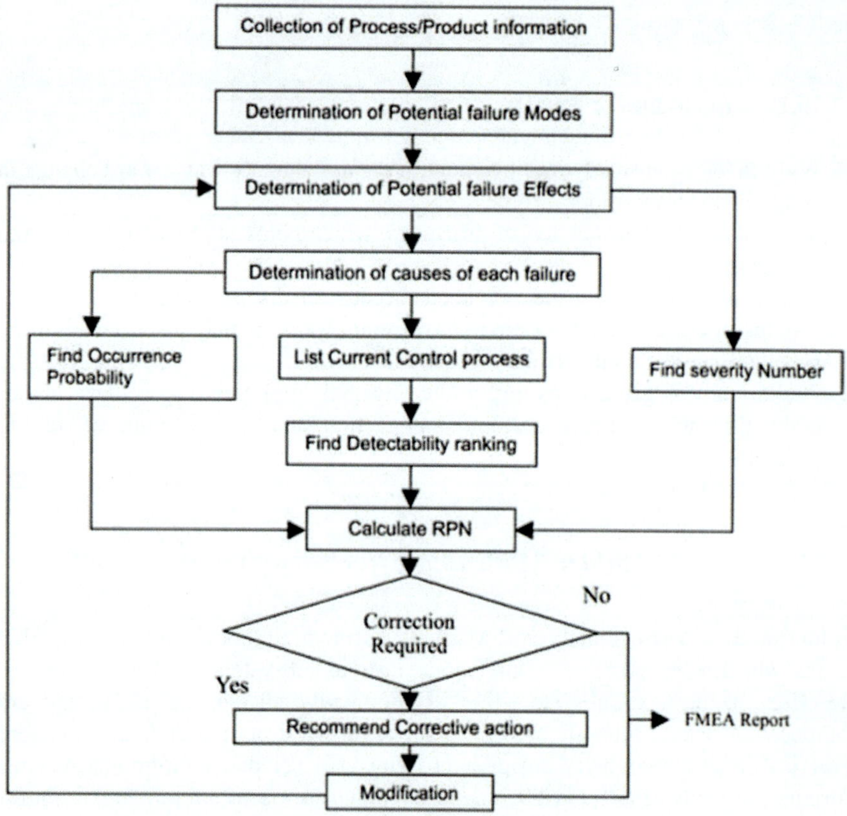

Fig. 2.8 FMEA flowchart

2.2.11.1 Step-Wise Procedure for FMEA

1. Determine the system to be analyzed. Further, divide the system into subsystems and/or assemblies in order to limit the search for components and prepare a list of components for each assembly.
2. In this step, build the block diagram of the system by using structural, functional or master logic diagram to recognize the relations among the subsystems/components.
3. Identify the possible failure modes of each component, their causes and the effects of failure modes on the immediate function or item, on subsystems and the entire system.
4. Perform assessment of each failure mode in terms of worst potential outcome (severity, S).
5. Find out the failure detection methods and compensating terms(s) for each failure mode.
6. Find the likelihood of the occurrence (S_f) of the failure mode using both qualitative and quantitative techniques.
7. Multiply S_f, S, S_d to obtain risk priority number (RPN).
8. In this step decide upon the remedial action depending upon the risk priority number. If remedial action is required than identify the design changes or other measures necessary to eliminate the causes of failure. These measures may be (i) compensatory (in order to minimize the loss in event of failure occurrence.) and (ii) Preventive to avoid recurrence failure situation.
9. Prepare recommendations to improve the system performance.
10. In the last step, develop FMEA report by summarizing the FMEA study in tabular format as presented in Table 2.4.

Table 2.5 presents simplified FMEA in an automobile assembly plant for seat belt installation process with three potential failure modes, i.e., selection of wrong color seat belt, seat belt bolt not tightened fully and trim cover clip misaligned. The second failure mode has an RPN of 144, followed by first failure mode. Thus, it receives the highest priority for process improvement.

Table 2.4 FMEA report

System											FMEA No.			
Subsystem Component Core team											Page Prepared by FMEA Date (org.)			
					Existing conditions					Action results				
Component/ process	Potential failure mode	Potential effects of mode	Potential causes of mode	Present control mechanisms	Severity	Occurrence	Detection	Risk priority number (RPN)	Recommend actions	Action taken	S	O	D	RPN

Table 2.5 FMEA example

Failure mode and effects analysis (FMEA) Process name: Left front seat belt installation				
Failure mode	(A) Severity (Rate 1–10)	(B) Probability of occurrence (Rate 1–10)	(C) Probability of detection (Rate 1–10)	Risk priority number
(1) Select wrong color seat belt	5	4	3	60
(2) Seat Belt bolt not fully tightened	9	2	8	144
(3) Trim cover Clip Misaligned	2	3	4	24

2.2.11.2 Advantages of FMEA

1. It adopts a well-documented procedure to select a design with a high likelihood of operational success.
2. It helps in the identification of critical areas of the system and thus helps in Fault/accident prevention by timely diagnosis.
3. It acts as a means for teamwork and idea exchange between functions.
4. It aims to improve the quality, reliability and safety of a product/process.
5. It reduces the possible warranty concerns and reduces late changes and cost associated with these changes.

2.2.12 Kanban

Kanban, a Japanese word given by Taiichi Ohno which means, "visual card." These cards were introduced and refined in Toyota to limit the inventory tied up in "work in progress" and to regulate the flow of parts in transfer lines in the 1950s. It starts with the customer's order and followed by production activities in downstream. It is frequently called a "pull" system because all the successive requests are pulled from the order.

The core concept of Kanban includes:

- **Visualize Workflow**: The first and foremost job is to identify the current flow of work, i.e., the sequence of steps to be executed to move an item from request to a deliverable product. Kanban board with cards is used to regulate the flow. The columns in the board represent a step in the workflow and each card represents a work item. By following this process, one can track the progress of work and identify blockages in real time.
- **Minimize Work in Process**: It limits work in progress as teams pay attention toward completion of all remaining work before initiating new work.

- **Manages flow of work**: It helps to manage the flow of work through the production process smoothly.

The commonly accepted rules for the success of kanban are:

I. The upstream processes should send items downstream in the precise quantities and order as specified in the Kanban card.
II. The downstream processes shall only take out items in the precise quantities as specified in the Kanban card.
III. A Kanban must accompany each item at all times. No items are permitted to be made or moved without a Kanban.
IV. Defects and imprecise quantities shall not be delivered to the downstream process in sequence.
V. It is important to monitor the number of Kanbans carefully to disclose the problems and opportunities for improvement.

2.2.12.1 Advantages of Kanban

Following are the advantages of Kanban:

- As blockages are clearly observed so, managers can work together to optimize the whole value chain rather than just their portion.
- Kanban is helpful for places where operations and supporting teams experience high incidence of uncertainty and variability.
- It helps to reduce inventory (25–75%) thereby lessening inventory holding costs. It eliminates waste of overproduction, thereby saving resources and time as well.
- It provides teams with sound principles for visualization of their work, which helps them to deliver the products and services in continuous manner and get customer responses with greater agility.

2.2.13 Poka-Yoke

Shigeo Shingo, an industrial engineer at Toyota found Poka Yoke (poh-kah yoh-keh) in Japan during the 1960s. He also formalized Zero Quality Control aimed at correcting the likely defects and do inspection to avoid defects. Poka Yoke is also translated as "resistance to errors," i.e., avoid (yoker) errors resulting from inattention (poka). Normally, the repetitive operations in a process depend on persons memory, and use of poka-yoke may save time and liberate the mind of workers for operations with more creativity. Today because of competition, it has become customary for the companies to produce products 100% defect free. To achieve this, Poka Yoke ensures that the accurate conditions be present before a process step is executed and thus aims to prevent defects from the first place.

Examples of Poka Yoke Application

One of the most common examples is when a driver of a car with manual gearbox must press on the clutch pedal (a process step—Poka Yoke) before starting the engine. The interlock prevents from an unplanned movement of the car. This serves as behavior-shaping constraint as there are actions that must be performed before the car is allowed to start. Other example may be the washing machine that does not start working if the door is open or not closed properly in order to prevent flooding. These types of automation in devices do not allow mistakes or incorrect operation.

2.2.13.1 Poka-Yoke Advantages

The method might be used on the occurrence of error. It can be fruitfully used to safeguard any type of process in manufacturing or services industry, thus avoiding all kinds of errors:

1. The technique is used to prevent processing error on account of missed operation or operation not performed as per the defined SOPs.
2. The technique is used to minimize setup error occurred because of the improper tooling or machine setups or adjustments.
3. The technique is used to identify the missing part for instance if all parts are not included in the assembly or other processes.
4. The technique is used to identify improper part/item, i.e., wrong part used in the process.
5. It identifies error if operations are carried out incorrectly.

2.2.14 Quality Function Deployment (QFD)

QFD is a methodology to listen the voice of the customer and then effectively responding to their needs. The tool was developed in Japan by Yoji Akao in 1966 while working for Mitsubishi's shipyard and was brought to the USA in the early 1980s. It gained its early appreciation as a result of successful application in the automotive sector. It is carried out by incorporating the needs of the end user, i.e., the customer into the required subsystems or during the phases of the product development ranging from design to development to engineering to manufacturing and finally to sales and services. The main justification for developing QFD is to venture a product to meet customer requirements.

According to the QFD Institute:

QFD is:

1. To understand the customer basic requirements.
2. To inculcate quality systems thinking + psychology + knowledge.
3. To maximize quality that adds value and eliminate nonvalue-added activities.

4. To adopt all-inclusive company-wide quality management system for end-user satisfaction.
5. To develop strategic plan to stay ahead of the competitors'.

QFD Methodology

Starting with the primary matrix, usually known as the House of Quality (HOQ) matrix, the QFD laid focus on the most significant attributes related to the product or service. The QFD process includes a series of matrices to represent data; the generally used matrix is the HOQ matrix (Hauser and Clausing 1988) as presented in Fig. 2.9. It represents the link between customer requirements and the technical or design requirements. With respect to product design, the "Whats" identifies the attributes related to the product or services most wanted by the customer whereas the "Hows" identifies the means to accomplish the "Whats." The central part of HOQ is called relationship matrix, is called the heart of the system, it presents the prioritization of the "Whats" by applying importance ratings. Importance rating is found by multiplying the importance level of each "What" (using a Likert scale) by its strength relationship with the "Hows," by making use of the symbols with corresponding weight values. Figure 2.10 shows the four matrices of QFD model, these matrices are used in each phase to translate customer requirements from the

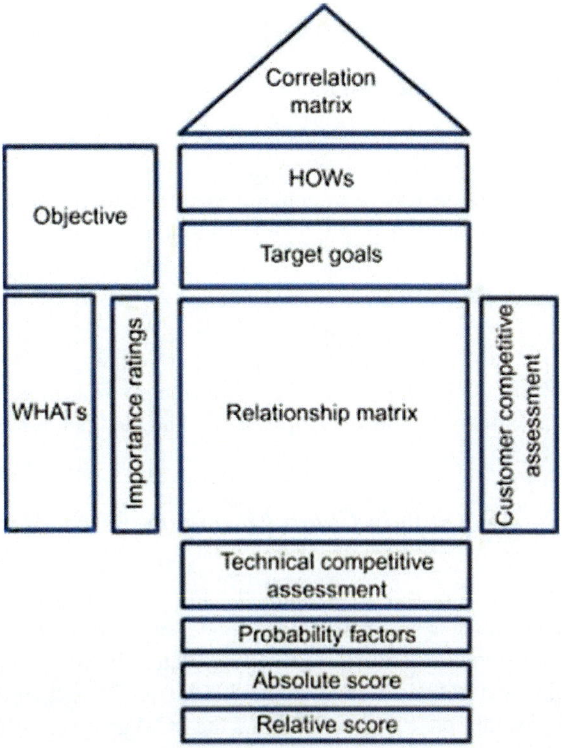

Fig. 2.9 HOQ structure (Hauser and Clausing 1988)

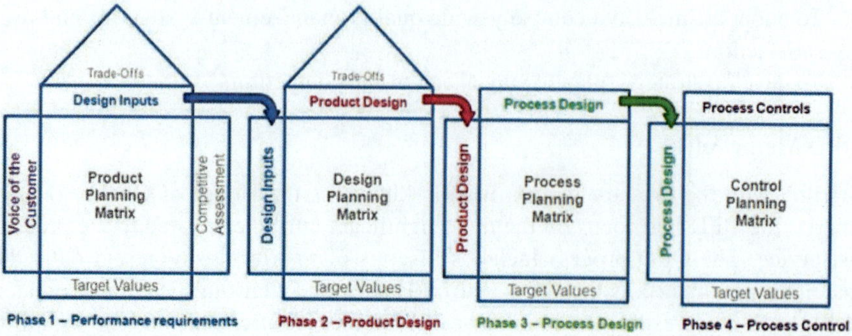

Fig. 2.10 QFD matrices

initial planning stages to process design to product design and finally process control. Each matrix consists of a vertical column called "WHATs" and a horizontal row called "HOWs." "WHATs" are customer requirements; "HOWs" are the possible ways called technical attributes to achieve them. At each stage, only the most significant aspects from "HOWs" are deployed into the next phase as "new WHATs" and the process continues until the last phase, i.e., process control.

2.2.14.1 Benefits of Quality Function Deployment

The possible benefits of QFD are as follows:

1. The rigorous approach adopted in QFD helps to draw the specification based on accurate set of requirements.
2. QFD ensures a systematic approach to the design process, which results in decreasing the design and manufacturing costs.
3. The method makes customer needs more visible by permitting an organization to prioritize and deliver them to the customers based on their requirements.
4. It aims to collect design data in a centralized format that is generally accessible to different teams working in their domains.
5. Better communication and information sharing among teams allow the identification of "holes" in the existing knowledge among them.

2.2.15 Supply Chain Management (SCM)

It is an approach to monitor the life cycle of materials as they flow in, through, and out of a business from production to distribution to returns. It involves the transportation and storage of raw materials, of work-in-process inventory, and of finished goods. It aims to meet order fulfillment from point of origin to point of consumption. Figure 2.11 shows the SCM process with systematic flow of materials,

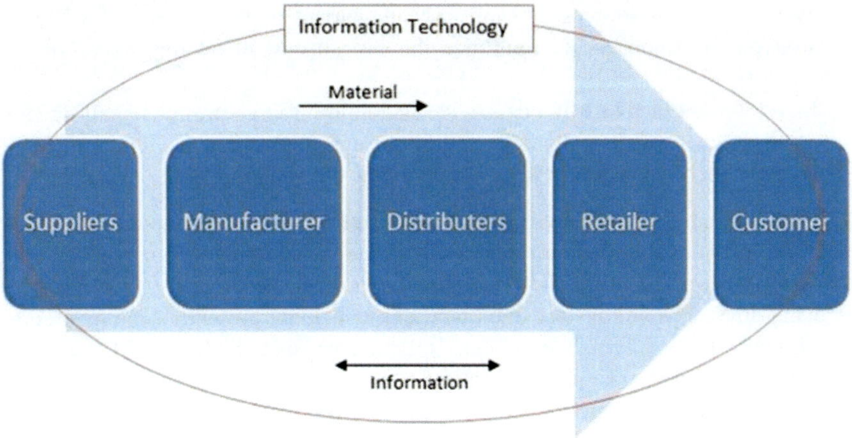

Fig. 2.11 SCM process among entities

goods, and related information among various entities, i.e., suppliers, manufacturers, distributors, retailers and consumers in a complex supply chain network.

In nutshell, a typical supply chain includes the following functions:

- Procurement.
- Sourcing of raw materials.
- Manufacturing and production activities.
- Shipment, receipts, storage, and fulfillment activities.
- Management of returns and exchanges through reverse logistics.
- Optimization of the general approach for profitability in supply chain.

According to supply chain experts (Anderson et al. 2007), the seven core principles for successful management of SCM activities are:

1. Segmentation of customers: Segment the customers based on their service requirements and acclimatize the supply chain process to handle each segment efficiently and profitably.
2. Customization: Customize the logistics network to meet the service requirements and profitability of each entity in supply chain network.
3. Development of demand forecasts: Development of demand forecasts should be done based on market demand across the supply chain for best possible resource allocation.
4. Product Differentiation: Distinguish the product in the supply chain with focus on downstream activities, because product differentiation helps in determination of true demand metrics. For instance, if a company has four clients who require four different versions of product being offered by the company, than as a manager, you shall forecast your demand based on number of core products that you will need for customer fulfillment.

5. Management of sources of supply: Management of the sources of supply is strategically important to minimize the overall cost of owning materials and services.
6. Development of a supply chain-wide technology: Organizations shall develop a supply chain-wide technology policy to support decision-making at different levels, which helps to provide a holistic view of products and services.

Planning and execution are important types of SCM software applications. The planning software applications make use highly developed algorithms to decide the best possible way to fulfill a product order. Since past few years, SAP has been focusing on SCM solutions provided by them, which are presently promoted under the Digital Supply Chain. The modern software products built on SAP Leonardo intelligent platform make use of machine learning and Internet of Things to bring visibility, forecasting, analytics, and partnership to the next level. Oracle SCM Cloud, JDA's Supply Chain Platform, Logility Voyager solutions provide innovative supply chain solutions for companies to manage their supply chain.

2.2.15.1 Benefits of Supply Chain

1. SCM process protects quality of the product and promotes development of sustainable products.
2. Responsible SCM processes document and comply with recognized CSR standards and guidelines internationally.
3. It improves association with suppliers and thus contributes toward productivity enhancements and reducing costs.
4. Companies with efficient SC process can overcome the bottlenecks to information flow among the various entities.
5. Efficient supply chains facilitate product flow among entities and also arrest the bullwhip effect in supply chain.

2.2.16 Six Sigma

Six Sigma (6σ) is structured methodology aimed to improve the process and increase customer satisfaction (both internal and external customers). It started at Motorola in the 1980s, and later on extended to AlliedSignal, General Electric and Honeywell International. The key notion behind Six Sigma is to decrease the variation in processes through continuous process improvement. Greek letter "Sigma" a statistical term; which is used to measure the deviation from actual target, for a given process. It permits an organization to determine the number of "defects" in a process, identify the ways to remove them and achieve "zero defects."

To accomplish 6σ targets, a process must not result in more than 3.4 DPMO. It achieves this goal by using DMAIC and DMADV processes. The acronym DMAIC stands for defines measure, analyze, improve, control. It is aimed for improving the

Fig. 2.12 6σ roles and responsibilities (*Image source* https://www.processexam.com/six-sigma-organization)

existing processes, which falls below the specification and are a candidate for incremental improvement. The acronym DMADV stands for define, measure, analyze, design, verify. It is aimed to develop new processes or products with desired six sigma levels. DMADV process can also be applied if the existing process requires more than just incremental improvement. Both DMAIC and DMADV processes are implemented by Green Belts (GBs) and Black Belts (BBs) and are supervised by Master Black Belts (MBBs). Figure 2.12 shows 6σ roles and responsibilities of key stakeholders.

Numerous quality tools can be used to support 6σ in an organization, such as Value Stream Mapping, Capability Analysis, 5 Whys, PDCA, and SPC. 6σ is not just applicable to manufacturing but can be useful to the financial services industry, says Doug Sutton, President Fidelity Wide Processing, where 6σ has brought cost reductions and improvements in quality in the range of 20–50%.

2.2.16.1 Benefits of Six Sigma

1. Six Sigma helps to improve business processes and sustain quality improvement efforts of organizations.
2. It works on proactive approach by identifying and providing solutions for likely problems impacting profitability of the company.

3. Six Sigma helps to convert the weaknesses into strengths by use of problem-solving tools.
4. Six Sigma delivers on both quality assurance and quality control activities and helps in successful implementation of quality management initiatives in organizations.

2.2.17 5S and House Keeping

5S system is a Japanese tool that aims to improve the workplace efficiency by eliminating the waste. There are five S in the system, each starts with the letter S, i.e., *Seiri, Seiton, Seiso, Seiketsu* and *Shitsuke*. In English, the five S's are means as Sort, Set in Order, Shine, Standardize and Sustain. Table 2.6 provides the description to these words.

2.2.17.1 5S and Housekeeping

The potential benefits from 5S systems are:

1. Under 5S, parts and tools are positioned in organized manner, which make them accessible and trouble free to operate.
2. 5S helps companies to reduce waste from the manufacturing and thereby decreasing the costs.
3. It helps in reductions in the required space and thus enhances storage density for existing operations.
4. 5S as part of visual control, identify the operational effectiveness by practicing discipline and promoting work culture.
5. It helps to master problem-solving skills and develop worker morale and enhances their engagement.
6. It aims to bring down the number of workplace injuries, thus reduces risk and improves safety.

2.3 Quality System Standards

The concept of quality has become a vital part of business in current times both at national level and global level. The quality management system (QMS) intends to improve organizational efficiency and increase customer satisfaction. QMS standards take into account eight principles, i.e., (i) customer focus, (ii) leadership, (iii) involvement of people, (iv) process approach, (v) system approach to management, (vi) continual improvement, (vii) factual approach to decision making and (viii)

Table 2.6 5S descriptions

Seiri (sort)	It aims to sort out or clear all of the waste from the workplace	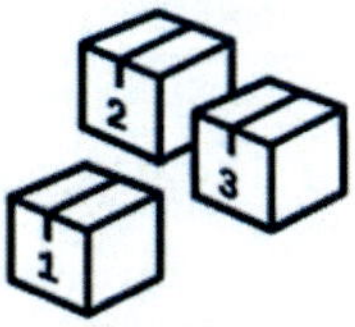
Seiton (order)	It means to organize all tools, materials and place them in order and ergonomic way to minimize the search	
Seiso (shine)	It means to cleanse the whole area to remove all dirt by sweeping	
Seiketsu (standardize)	It aims to standardize the work practices and provide stability to operations in the first three stages	
Shitsuke (sustain)	It means to ensure that 5S principles are part of the business culture and practice	

mutually beneficial supplier relationships. The adoption of these standards enhances customer satisfaction by meeting customer requirements.

2.3.1 ISO 9000 and Other Derivatives

The International Organization for Standardization (ISO) currently has a collection of over 21,000 standards, of which the ISO 9000 series relating to quality management is certainly the best known. Within this series, ISO 9001 (QMS requirements) has been extensively used by organizations around the world to provide evidence that they have a clearly defined and well-managed set of processes that facilitate them to constantly supply products and services meeting customer and appropriate constitutional and regulatory requirements. The primary set of standards has undergone several revisions from time to time.

ISO 9000—Quality Management and Quality Assurance Standards—guidelines for selection and use;

ISO 9001—Quality Systems—Model for Quality Assurance in design/development, production, installation and servicing;

ISO 9002—Quality Systems—Model for Quality Assurance in Production and Installation;

ISO 9003—Quality Systems—Model for Quality Assurance in Final Inspection and Test;

ISO 9004—Quality Management and Quality System Elements—GUIDELINES. Supporting Standards on Vocabulary/Terminology;

ISO 8402 (1986) IS 13,999 now IS 10,201 Part I (1988).

ISO 9000 Certification: The worldwide importance on quality has resulted in many enterprises to evaluate their suppliers' processes. These audits of supplier capabilities are occasionally a precondition to purchase activities. Enterprises have learnt that the adoption of good quality processes produces superior products. In year 2000, when the standards were revised for the second time, the title Model for a Quality Assurance System was replaced by Requirements for a Quality Management System. All the three standards 9001, 9002 and 9003 were merged into one standard, i.e., ISO 9001. Further, the 2008 revision of ISO 9001 was a big revision with respect to the contents and presentation. The revised 2008 standards were based on eight basic management principles, as shown in Table 2.7 with description.

Other Industry Standards: Various industries are adopting to standards, similar to ISO 9000, but modified to meet their specific needs. Table 2.8 presents the comparison of few of these standards (source: https://tl9000.org/about/tl9000/table.html).

Table 2.7 Eight basic management principles

Principle	Description
Customer focus	Helps to increase revenue and market share through flexible and fast responses to market opportunities
Leadership	It offers unified way to align, implement and evaluate the activities, thereby motivating the people to understand the organizational goals and objectives
Involvement of people	It helps the people to participate and contribute toward continuous improvement
Process approach	It provides focused and prioritized improvement initiatives, which help to reduce costs and reduce cycle times through efficient use of resources
System approach to management	It integrates and aligns the key business processes to achieve the desired results
Continual improvement	It provides flexibility to react quickly to opportunities and hence improves business performance through improved managerial abilities
Factual approach to decision-making	It helps in making informed decisions, which is possible with increase in ability to review, challenge and amend decisions
Mutually beneficial supplier relationships	With increased ability to create value for both parties, it helps to build mutually beneficial supplier relationships

2.4 Action Plan for Implementing ISO 9001:2008 Standards

Table 2.9 presents the steps for ISO 9001 implementation in an organization.

2.5 Concluding Remarks

This chapter explained various quality management tools, which are necessary to design, develop and analyze quality problems in order to meet quality standards with respect to products and services and satisfy ever-growing customer needs. The tools like Check sheet, 5 S, Kaizen, Scatter diagram, Histogram, Pareto Diagram are easy to use and can be adopted for getting effective results whereas other tools such as FMEA, Six Sigma, Kanban, QFD, CRM need extensive training and expertise, with proper management support/commitment for effective implementation. Also, a discussion on Quality System standards with implementation steps is provided.

Table 2.8 Industry standards

	TL 9000	TS 16,949	CMMI
Description	The standard adds detailed telecom requirements and measurements to standard ISO 9000 practices	It adds defined automotive requirements to generic ISO 9000 practices	It describes standard best practices for creating systems in software engineering domain
Requirements	It comprises of specific and detailed requirements as related to telecom sector	It comprises of specific and detailed requirements related to automotive sector	It requires that targets to be met, expects that practices related to those goals are implemented and provides suggestions for detailed implementation of the practices
Measurements	It has a very strong focus on well-defined postdeployment measurements	It requires the use of measurements tied to business objectives for the purpose of continual improvement	It focuses primarily on predeployment best practices and in-process measurements
Business and/or manufacturing focus	It incorporates business needs as specific requirements for hardware, software and services development	It incorporates business needs with an emphasis on hardware development and manufacturing	It requires that all the process areas, goals, and practices be interpreted based on business needs of the organization
Role of customer	It defines specific customer inputs and involvement as stakeholders	Not specified	It expects the organization to identify and involve relevant stakeholders but does not specify them

Table 2.9 Steps for ISO 9001 implementation

The required steps for ISO 9001 implementation in an organization consist of:	
Step 1	Constitution of Management Committee
Step 2	Defining the scope
Step 3	Present organizational structure
Step 4	Perform review and finalize Vision and Mission statement
Step 5	Identification of processes, along with designations
Step 6	Defining the sequence and interactions among the processes
Step 7	Provide explanation of duties of each designation
Step 8	Preparation of the list of documents

(continued)

Table 2.9 (continued)

(1) Make presentation to the leadership on the new requirements of the standard
(2) Set up a steering committee for the changeover to ISO 9001:2015
(3) Perform a Gap Analysis
(4) Determine/explain background of the organization
• Identification of internal/external issues
• Conduct SWOT/PEST/LE analysis
(5) Determination of interested parties and their needs and expectations
(6) Performing review and the scope of the QMS
(7) Performing review of the processes of the organization, and ensure that they must consist of the required information for the process approach
(8) Performing review of quality policy to ensure alignment with context and strategic direction
(9) Identifying risks and future prospects throughout the organization for all key processes/projects related to QMS implementation
(10) Performing review of quality objectives for proper alignment with company policy, so as to ensure that they are measured, evaluated and there is an action plan to achieve them
(11) Considering developing a communication plan for all key communications in the organization
(12) Communication of the requirements of ISO 9001:2015 to the entire organizational units. Ensure that all employees understand:
• the quality policy;
• the quality objectives relevant to them;
• their contribution to the effectiveness of the QMS and benefits of improved performance; and
• Implications of not conforming to the QMS requirements
(13) Make sure that there is a system for addressing change management
(14) Determine/clarify the systems for capturing and maintaining organizational knowledge
(15) Conduct review monitoring, measurement, analysis and evaluation, including customer satisfaction
(16) Conduct internal audits to the new version of the standard
(17) BRING ON THE REGISTRAR!!

References

Anderson DL, Britt FF, Favre DJ (1997) The seven principles of supply chain management, logistics management. Supply Chain Manag Rev, Conference, Thailand, 2007

Brkic VK, Djurduja T, Dondur N, Klarin MM, Tomic B (2013) An empirical examination of the impact of quality tools application on business performance: evidence from Serbia. Total Qual Manag 24(5):607–618

Chambers JM, Cleveland WS, Kleiner B, Tukey PA (1983) Graphical methods for data analysis. Wadsworth, Belmont, CA

Cristiano JJ, Liker JK, White CC III (2001) Key factors in the successful application of quality function deployment (QFD). IEEE Trans Eng Manag 48(1):81–95

Crosby PB (1979) Quality is free. McGraw-Hill, New York

Dale BG, McQuater R (1998) Measuring business improvement and quality: implementation key tools and technique. Blackwell Business, Oxford

Gartner: market share: customer experience and relationship management, worldwide, 2018

Garvin DA (1984) What does product quality really mean? Sloan Manag Rev 26(1):25–43

Goetsch D, Davis S (2003) Overview of total quality tools. Quality management: introduction to total quality management for production. Processing and service, Prentice-Hall, Upper Saddle River, NJ

Harland CM (1996) Supply chain management, purchasing and supply management, logistics, vertical integration, materials management and supply chain dynamics. In: Slack N (ed) Blackwell encyclopedic dictionary of operations management. Blackwell, UK

Hauser JR, Clausing DP (1988) The house of quality. Harv Bus Rev

Ishikawa K (1976) Guide to quality control. Asian Productivity Organization, Tokyo

Juran JM (ed) (1974) Quality control handbook, 3rd edn. McGraw-Hill, New York

Juran JM, Gryna FM Jr (1993) Quality planning and analysis: from product development through use, 3rd edn. McGraw Hill, New York

Maritan D (2014) Quality function deployment (QFD): definitions, history and models. Practical manual of quality function deployment, 1–32https://doi.org/10.1007/978-3-319-08521-0

McQuarter RE, Scurr CH, Dale BG, Hillman PG (1995) Using quality tools and technique successfully. TQM Mag 7(6):37–42

Sharma RK, Kumar D, Kumar P (2007) Quality costing in process industries through QCAS: a practical case. Int J Prod Res 45(15):3381–3403

Talib F, Rahman Z, Qureshi MN (2013) An empirical investigation of relationship between total quality management practices and quality performance in Indian service companies. Int J Qual Reliab Manag 30(3):280–318

Tari JJ, Sabater V (2004) Quality tools and techniques: are they necessary for quality management? Int J Prod Econ 92(3):267–280

Chapter 3
Quality Management in MSME Sectors

Highlights

To understand the importance of quality management practices for MSMEs, *this chapter provides details to*

- Problem definition and Study objectives
- Research methodology and sample size determination
- Design of survey instrument for accomplishing the objectives.

3.1 Introduction

Quality management practices have been widely used by large business houses such as Hero, Tata Motors, Maruti Suzuki, IOCL, etc., but their adoption and practices in MSME sector are not adequate, which calls for necessary attention by the researchers in this field. There is no doubt that Indian MSMEs are passing through a transitional period of high market competition with escalating demands of consumers for getting better product and service (Subrahmanya 2005). For the survival of MSMEs in this ever-expanding marketplace, customer satisfaction, improvement in productivity, changing organizational culture and globalization of world trade, the emergence of quality plays vital role (Zakun et al. 2010). Quality is widely recognized as one of the most important discipline/strategies or competitive priority for an organizational development (Sharma and Kodali 2008). This is why worldwide many companies have become aware and understand the need to make quality as their marketing strategy in a global market.

R .K. Sharma, *Quality Management Practices in MSME Sectors*,
Springer Tracts in Mechanical Engineering,
https://doi.org/10.1007/978-981-15-9512-7_3

This chapter basically tries to investigate the role of quality management in competitive positioning of MSMEs. To this effect, a structured questionnaire is prepared and MSMEs situated in state of H.P are considered. Aim was to get maximum response from industries so data were collected mostly by making personal visits, only few through email, online questionnaire, and postal service.

3.2 Problem Definition

The concept of quality management is new to the Indian MSMEs. Even though, the ISO certification of quality standard was launched during 1990, MSMEs are still in process to adopt these standards. Most of them are just satisfied with the level of quality products they are manufacturing. However, in current high competitive ever changing market scenario, they are losing market share to competitors. They have felt the importance of growth, customer satisfaction and employee empowerment. Some MSMEs have begun to adopt quality management practices as an approach to fulfill these objectives. However, there is a strong need to understand the importance of quality management practices for MSMEs, so as they can improve their current business climate through improved quality of product and service, and ensure their long term survival. Furthermore, as far Indian industries concern, their focus primarily on ISO 9000 certification, and many of them not succeed to fulfill these criteria. MSMEs are not even thinking beyond ISO certification for them QM is far to achieve. In light of this situation, there is need for a deeper investigation of the relationship between QM practices and MSMEs performance and competitive positioning. It is to fill this research gap that the present study is being undertaken. The present study covered MSMEs in Himachal Pradesh, India. Figure 3.1 shows the conceptual framework described to conduct an in-depth study. The key contents of the frameworks are:

i. Clarification of context
ii. Evaluating the main constructs and issues
iii. Questionnaire development
iv. Identifying target population of MSMEs and sample size determination
v. Respondent's survey
vi. Data Collection and filtration
vii. Section-wise data analysis and discussions of results.

3.2.1 Study Objectives

The specific objective of this study comprised of the following:

i. To study the enthusiasm of adopting quality management practices among MSMEs.

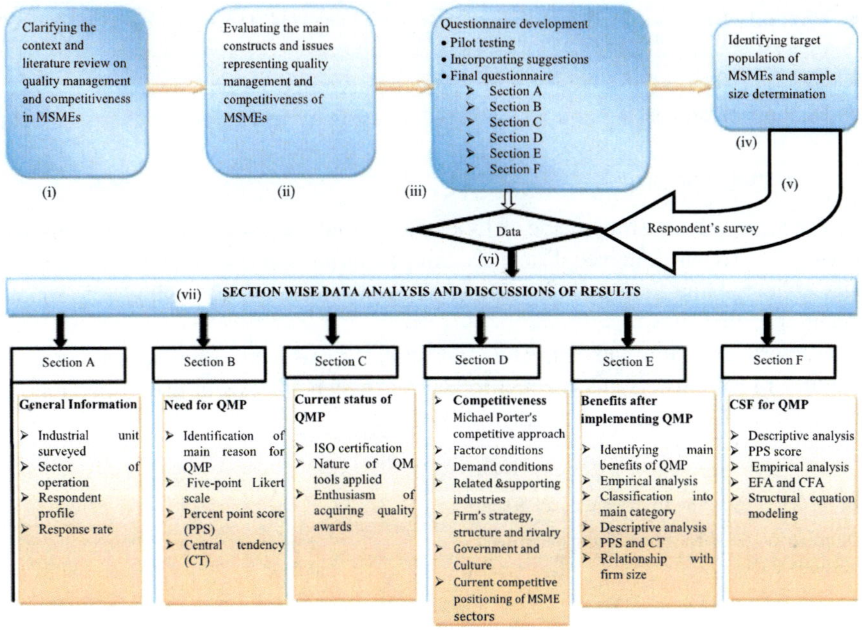

Fig. 3.1 The conceptual framework

ii. Collection of useful data pertaining to awareness regarding quality management.
iii. Categorizing ISO certified MSMEs on the basis of pollution control board norms
iv. To assess the willingness of SMEs to develop their systems of quality and enhance their level of quality standards such as ISO 9000
v. To study in which MSME sectors quality management practices are applied.
vi. To study the applicability of various quality tools in MSMEs and stages in company where quality management practices are applied.
vii. To measure competitive positioning in MSMEs using Porter's Diamond model indicators.
viii. To study benefits obtained from implementing quality management practices in MSMEs.
ix. To identify critical success factors that may help in successful adoption of quality management practices.

3.2.2 *Research Methodology*

Research methodology adopted in the study is basically survey based upon a well-designed questionnaire.

- Method of data collection = Through Questionnaires, Interviews and Industrial visits.
- Reference period of the data to be covered = Around 1 year
- Method of processing and analyzing—Analysis by means of pie charts and bar charts
- Case study approach.

For survey administration, the method of the postal survey followed by e-mails and personal visits was followed. Data pertaining to various sections in the questionnaire was collected and Percent Point Score (PPS) is calculated, according to formulae given in Table 3.1.

As per discussion with Director, Industries Department, Shimla, Govt of H.P, the four major clusters are selected for survey, which consists of firms operating in different MSMEs sectors in the region (Fig. 3.2).

Table 3.1 Procedure to determine PPS score

Number of responses (N)	Number of units scoring					Total Point Score (TPS)[a]	Percent Point Score (PPS)
	1 (W_1)	2 (W_2)	3 (W_3)	4 (W_4)	5 (W_5)		$\frac{TPS}{5*N} = 100$

[a]Total Point Score (TPS) = $1 \times W_1 + 2\times W_2 + 3\times W_3 + 4\times W_4 + 5\times W_5$

Source Nanda and Singh (2009)

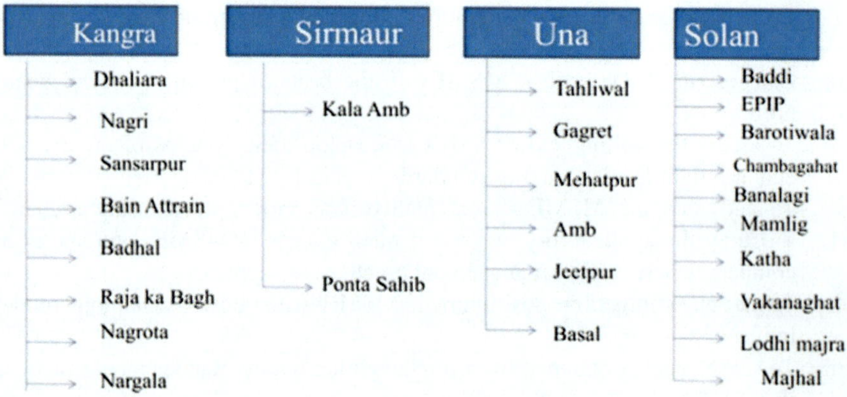

Fig. 3.2 Industrial clusters

3.2.3 Sample Size Determination

After problem definition, the next step is to determine appropriate sample size. According to the Director Industries, Industries Department, Government of Himachal Pradesh, the State has total 39512 units in the small-, medium- and large-scale sector, which comprises of various sectors, i.e., Pharmaceutical and allied, Food and allied, Mechanical, Electrical and Electronics, Textile, Paper and Paper Products. The detail of location of industries spread over different districts of Himachal Pradesh is presented in Table 3.2. Figure 3.3 presents the distribution of nature of Industry according to product type in state. Target population of MSMEs considered in the study is 39018 units (as per official website of Industries department, H.P Govt.) working in a variety of areas (*metal products, precision parts, pharmaceuticals, wood products, textiles, electronics, healthcare, food beverage, paper and printing, chemical, machinery and automotive engineering units)* spread across various districts, SHIMLA, UNA, SIRMAUR, SOLAN, BILASPUR, HAMIRPUR and KANGRA of Himachal Pradesh as shown in Table 3.3.

Table 3.2 Location-wise details of industry

Name of location	Primary industry	Description
Baddi	Textiles	Located in the Solan District. Pharmaceutical, Textile, Auto Ancillary, Paper and Printing Units
Barotiwala	Multiindustry	Located in the Solan District Hub for small industries
Parwanoo	Light Engineering	Located in the Solan district: An industrial cluster for high density polyethylene (HDPE) pipe companies
Chambaghat	Electronics	Located on the Shimla-Kalka highway in the Solan district, the industrial area has prominent companies such as Himachal Futuristic Communication Limited and Shivalik Bimetal and controls Limited
Ponta Sahib	Pharmaceutical/Chemicals	Located in Sirmour district. The industrial area has a number of pharmaceuticals and chemical companies
Tahliwal and Gagret	Electronics	Located in the Una district, it has prominent telecommunication companies
Amb Industrial Area	Engineering/Auto	Amb has prominent engineering and automotive companies such as ICML
Sansarpur Terrace	Engineering	Located in the Kangra district, it has some of the leading engineering companies in North India

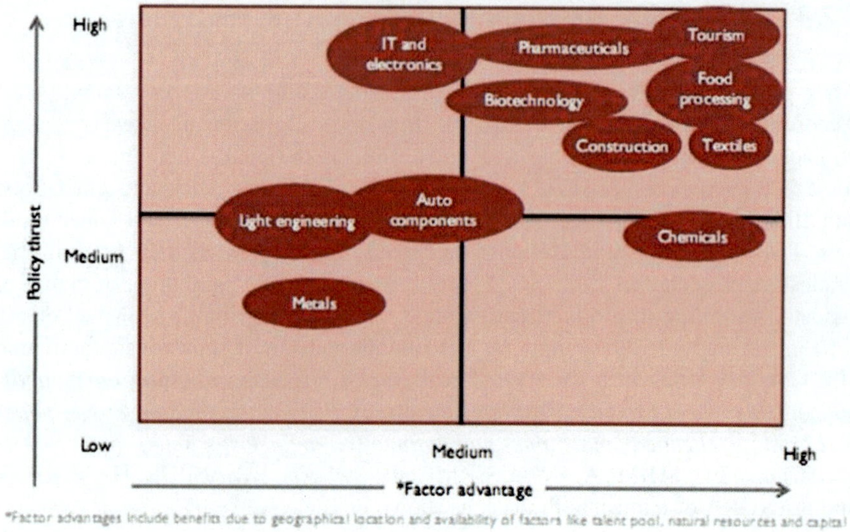

Fig. 3.3 Distribution of nature of Industry (according to product type) (*Source* Report by IBEF, www.ibef.org STATE ECONOMY AND SOCIO-ECONOMIC PROFILE Himachal Pradesh November 2010)

Table 3.3 Year-wise details of MSMEs units in H.P

Sr. number	Year	Number of units set up	Investment (Rs. in lakhs)	Employment generated
1	up to 2002–2003	30176	70977.48	129871
2	2003–2004	663	3708.48	3769
3	2004–2005	913	8891.44	6412
4	2005–2006	914	12217.3	6611
5	2006–2007	952	45272.78	10665
6	2007–2008	842	70637.33	11302
7	2008–2009	909	73795.48	10939
8	2009–2010	1032	75320.01	10011
9	2010–2011	963	96539.36	10002
10	2011–2012	856	38584.00	7732
11	2012–2013	798	96331071	9298
	Total	**39018**	**457359.66**	**199582**

Source admis.hp.nic.in/himachal/industry/welcomelat.htm

According to Cochran Formula (Eq. 3.1), the sample size is = n,

$$n = \frac{n_0}{1 + \frac{n_{0-1}}{N}} \tag{3.1}$$

$$n_0 = \frac{Z^2 p.q}{C^2} \tag{3.2}$$

where: Z = value of confidence level at 95% (1.96) in normal distribution
C = confidence interval (5%)
P = estimated proportion of an attribute that is present in the population (0.5), in other words it is the probability that a particular observation will be selected in the sample, in the worst case it is 0.5 (50%). meaning every observation has an equal chance of being selected in the sample q = (1 − p)

Now, Calculating sample size according to finite population,

Where N = (population, i.e., no of MSME units in the state, i.e., N = 39018) We get, n = 379

This means a survey 379 units of the H.P state located in various districts. Further, in order to determine that 379 units to be surveyed, we have to determine the number of units group or sector wise. For this purpose based upon the information available in literature, the proportions of industries from one particular group are estimated as: 379/39018 = 0.0097.

Table 3.4 Number of units surveyed

MSME Sectors	Sector-wise units to be surveyed	Sector-wise response obtained
Food and allied	44	34
Textile	18	09
Wood and wood products	08	03
Paper and paper product	38	23
Leather and Lather products	10	06
Glass and ceramic	11	06
Chemical and allied (Pharmaceutical)	88	79
Mechanical Items	78	60
Electric and Electronic	70	60
Misc and other	14	06
Total	379	286

To get the required sample size of particular group, the actual number of units is multiplied by the proportion. Table 3.4 shows the respective number of units surveyed.

3.2.4 Design of Pilot Instrument

The survey instrument was designed to investigate level of quality practices in MSMEs. Due consideration was given to make the questions clear and straightforward based upon common quality practices and tools with simple language. Further, the questionnaire was designed in a simple tabular format so that respondents only tick their responses in the appropriate box. The questionnaire was divided into six sections marked as section A, B, C, D, E and F. Section A keeps track on the general information of responding companies and details of person filling questionnaire. In this section, background aspect of the responding company such as type, category as per pollution control board, year of establishment and awareness about quality and quality system certification were included in this section. In section B, we used five-point Likert scale to collect information pertaining to reason for introducing quality management practices and gaining quality standard certification. Section C contained information about the nature of quality management tools applied (cause and effect diagram, flow chart, control charts, six sigma, etc.), stages where they applied and year of adoption of these quality practices. Section D of the questionnaire deals with questions related to various variables (based upon Porter's diamond model) for assessing competitiveness in MSMEs. Section E presents questions on five-point scale to assess the company's experience after implementing quality management practices. Finally, the last section F presents questions pertaining to various critical success factors, which can help MSMEs in adoption of quality practices.

3.2.4.1 General Information and Respondents Profile

This section was designed to get general information about companies and respondents profile. It represents the questions about (i) the type of business, i.e., Private Ltd., a government enterprise, joint venture or a company operating in partnership, (ii)year of establishment and category type, i.e., micro, small and medium enterprises and (iii) sectors of operation, i.e., mechanical, electrical, electronics and chemical, etc. Furthermore, the section also represents the respondent's profile, which includes their work profile, i.e., CEO, plant manager, production manager, quality manager, etc., their working experience and educational detail. Annexure 1 presents the details of the questions regarding the general information.

3.2.4.2 Need for Quality Management Practices and Quality Standard Certification

The aim of this section is to highlight the possible reasons behind moving toward QM practices. Based on the extent review of literature, the leading causes of implementing QM practices can be explained by various quality dimensions, such as (i) to increase quality consciousness among the employees, (ii) to adopt a formalized system that documents the structure, responsibilities and procedures for effective quality management, (iii) to be competitive and successful in business operations both domestic and foreign market, (iv) to meet customer requirements and ensure consistency in products and service delivered, (v) to identify inefficient processes and promote lean manufacturing, (vi) to withstand pressure from existing supplier(s), (vii) to reduce potential audit expenditure both internal as well as external, (viii) to reduce warranty and replacement expenditures, (ix) to reduce the quality costs i.e. prevention, appraisal and failure costs, (x) to upgrade skills/training of employees, (xi) to increase productivity, (xii) to increase profitability, (xiii) to develop a nonpolluting environment-friendly manufacturing process with minimal waste, (xiv) to reduce potential energy expenditure and (xv) to sustain long-term growth, dynamism and employment opportunities. The survey questionnaire for this section was designed using this dimension. Further, five-point Likert scale representing 1 = not important, 2 = somewhat important, 3 = moderately important, 4 = important and 5 = very important is used to obtain respondents choices. The suggestions provided by renowned experts for this section were incorporated, which were mainly related to the wording of issues, structure and order of questions. All suggestions were incorporated before the actual field run of the survey instrument. Annexure 2 presents the details of the questions regarding the quality management practices and quality standard certification.

3.2.4.3 Information on Quality Management Practices

This section presents details with respect to:

i. ISO certification of firms
ii. Effect of ISO 9000 on the implementation of quality management practices
iii. Enthusiasm for acquiring quality awards
iv. Nature of tools and technique being implemented by the firms
v. Stages in a company where these tools are being used, i.e., incoming, work in progress and finished product stage. Annexure 3 presents the details of the questions regarding the points discussed above.

3.2.4.4 Section D: Information on Competitiveness

This section is based on Porter's diamond model and its determinants. They are:

- Factor conditions;
- Demand conditions;
- Related and supporting industries;
- Firm's strategy, structure and rivalry; and
- Role of Government.

The survey instrument (as shown in Annexure 4) for measuring these determinants consists of 55 questions, thus covering all influencing factors, which may affect competitiveness in MSMEs. The drafted instrument was sent to five (three academics and two industrial) practitioners, working in this field. Based upon their suggestions, items were modified, and total numbers of questions were reduced to 49. Further, pilot testing of the instrument was accomplished, and suggestions from 15 recognized MSMEs were incorporated. Items of the entire construct were measured as: (1) strongly disagree, (2) disagree, (3) neutral, (4) agree, (5) strongly agree for the first four determinants and (−2) significantly disappointed, (−1) disappointed, (0) natural, (+1) appreciable, (+2) for significantly appreciable for other construct's items, i.e., the role of government toward competitive positioning of MSMEs.

3.2.4.5 Information on Benefits After Implementation of QM Practices

The aim of this section is to identify the potential benefits after implementing QM practices. For this purpose, extent review of the literature was carried out, and 16 potential benefits were identified. These variables have been incredibly discussed in previous studies namely they are: (i) employees participation, (ii) employee training and education, (iii) employees morale, (iv) competitive spirit, (v) rewards and reorganization, (vi) recruitment and retaining of skilled men power, (vii) manufacturing lead time, (viii)energy consumption, (ix) maintenance cost, (x) rework and scrap, (xi) warranty expenditure, (xii) sales volume, (xiii) customer complaints, (xiv) company profit, (xv) customer relationships and (xvi) suppliers relationships.

The survey instrument (as shown in Annexure 5) for collecting primary data was designed and discussed with the group of experts from industries and academia to highlight the major issues of concern. The questions framed for this section represent portions as 2 = substantially increase, 1 = increase, 0 = no change, −1 = somewhat reduced and −2 = substantially reduce in possible benefits after implementing QM practices. Further, pilot testing of the instrument was accomplished, and suggestions from 15 recognized ISO 9000 certified MSMEs were incorporated.

3.2.4.6 Critical Success Factors in Adoption of QMP

This section of survey questionnaire (as shown in Annexure 6) was designed using extant review of literature thus covering all questions, which may influence, which may help MSMEs in adoption of Quality Management Practices. The work provided by Prajoga (2005), Ooi et al. (2012), Tari (2005), Tari et al. (2007), Talib et al. (2013) forms the basis for the questionnaire. The measurement of items related to QM practices and its effects on firm performance, however, required a more in-depth examination. The draft instrument was sent to five (three academics and two industrial) practitioners working in this field. Inputs from academicians, consultants and professionals from industries were used to modify it.

3.2.5 Pretesting and Validation of Questionnaire

After framing the questionnaire, the same was sent to eminent experts in this field so that the different sections pertaining to study can be critically examined. After then, the pilot test of the questionnaire was performed by a personal visit to organizations; about seven respondents have been approached to evaluate the validity of the questionnaire.

3.2.6 Respondents and Their Profile

CEO or Top management representatives are considered as respondents for making personal visits and discussions for collection of necessary inputs with respect to questionnaire sections. Based upon the sample size, survey was carried out in four main industrial clusters of the state, i.e., ***Kangra, Solan, Sirmour*** and ***Una***. The survey covered 379 MSME units in above mentioned clusters. Out of the total 379 questionnaires administered through postal, e-mail and personal visits, 286 were found appropriate, thus yielding a response rate of 76.26%, which is the sufficient number for further analysis.

3.3 Concluding Remarks

This chapter has described in detail the problem definition, population and sample size, research methodology for the study along with details of industrial clusters for survey administration. The next chapter presents the findings of the study.

Annexure 1: General Information

Q.I Name of the company...

Postal Address……………………………………………………………………………….....

....

……………………………………………………………………….......................

Pin Code: ☐☐☐☐☐☐

Phone Number (With STD Code)												
Fax Number With Code												
Website:												

Q.II Details of the contact person filling questionnaire.

Name and Designation……………………………………………………………………....................

Phone Number (With STD Code)												
Mobile Number												
Email Address:												

Q.III Company Type and Year of establishment Please tick (.√) appropriate Box.

☐ Private Ltd. ☐ Government ☐ Joint Venture ☐ Others

Year of Establishment: ………………………………………………………………………..

Q.IV Category of firm

☐ Micro Scale Unit ☐ Small Scale Unit ☐ Medium Scale Unit

Q.V Category of firm on the basis of the Pollution Control Board Norms

☐ Green ☐ Orange ☐ Red

Q.VI Major Product Manufactured. (Give name(s) of two Products at least) During last three years

1. …………………………………….. 2. ……………………………………

Q.VII Are you aware about the quality? (In terms of product quality)

☐ Yes ☐ No

Q.VIII Whether your company is ISO certified?

☐ Yes ☐ No

Annexure 2: Information on Need of Quality Management Practices and Quality Standard Certification

Please tick the appropriate box representing the reasons for introducing a quality management system in your company and gaining quality standard certification against the options on the scale 1–5; 1 = Not important, 2 = Somewhat important, 3 = Moderately important, 4 = Important, 5 = Very important.

S. No.	Reasons	1	2	3	4	5
1.	To increase quality consciousness among the employees					
2.	To adopt a formalized system that documents the structure, responsibilities and procedures for effective quality management.					
3.	To be competitive and successful in business operations both in domestic and foreign market.					
4.	To meet customer requirements and ensure consistency in products and service delivered.					
5.	To identify inefficient processes and promote lean manufacturing.					
6.	To withstand pressure from existing supplier(s).					
7.	To reduce potential audit expenditure both internal as well as external.					
8.	To reduce warranty and replacement expenditures.					
9.	To reduce the quality costs, i.e., prevention, appraisal and failure costs.					
10.	To upgrade skills/training of employees.					
11.	To increase productivity.					
12.	To increase profitability.					
13.	To develop a non polluting environment-friendly manufacturing process with minimal waste.					
14.	To reduce potential energy expenditure.					
15.	To sustain long-term growth, dynamism and employment opportunities.					

Annexure 3: Information on Quality Management Practices

Q.I. Sectors of operation. Please Tick (√) the appropriate box against each item

1. Food and allied ☐
2. Textile ☐
3. Wood and wood products ☐
4. Paper and paper product ☐
5. Leather and leather product ☐
6. Glass and ceramic ☐
7. Chemical and allied (Pharmaceutical) ☐
8. Mechanical items ☐
9. Electric and Electronic ☐
10. Misc and others (pl specify) ☐

Q.II. Year-wise Adoption of Quality Practices/Programs. Please (√) the appropriate box

1. Before-1990
2. 1991-1995 ☐
3. 1996-2000 ☐
4. 2001-2005
5. 2005 -2010 ☐
6. 2010-ownwards ☐

Q.III. Stages in Company where quality management practices are applied. Tick (√) more than one option, if applicable)

1. Incoming raw material/Vendors Inspection ☐
2. In process Inspection ☐
3. Finished Product Inspection ☐
4. Packaging ☐
5. After Sales Service ☐
6. Others (pl specify) ☐

Q.IV. Nature of Quality management tools applied. Tick (√) more than one option, if applicable)

1. Cause & Effect Diagram ☐
2. Customer Relationship Management (CRM) ☐
3. Check Sheets ☐
4. Control charts ☐
5. Enterprises Resource Planning (ERP) ☐
6. Kaizen ☐
7. Failure Mode Effect Analysis (FMEA) ☐
8. Histogram ☐
9. Kanban ☐
10. Pareto Diagram ☐
11. Process Flow Diagram ☐
12. Poka-yoke ☐
13. Supply Chain Management (SCM) ☐
14. Scatter Diagram ☐
15. Quality Function Deployment ☐
16. Six Sigma ☐
17. 5S and House Keeping ☐
18. Others * ☐

Note: * "Others" in Q.IV includes Total employee involvement (TEI), Statistical process Control (SPC), Total preventive maintenance (TPM), Quality Circles, Brain storming.

Annexure 4: Information on Competitiveness

Please tick the appropriate boxes representing variables to measure competitive positioning in your company on 1–5 scale with 1—Strongly disadvantage, 2—disadvantage, 3—Neither advantage nor disadvantage, 4—advantage and 5—strongly advantage.

Determinants	Casual variables	Proxy variables	1	2	3	4	5
Factor conditions	Basic factors	Natural resources					
		Physical resources					
		Unskilled labor					
		Highly educated personnel					
	Advanced factors	Production and process technologies					
		IPR and Patents					
		Communication infrastructure					
Demand conditions	Market Volume	Market size					
		Pattern of growth					
	Sophistication	Distribution channels					
		New investment in region					
Related and supporting Industries	Related companies	Technology upgrading,					
		information flow					
		Shared technology development					
	Support	R&D investments					
		Suppliers and distribution channels					
		Marketing					
Firm structure strategy, and rivalry	Rivalry	Boosted innovation					
	Structure/strategy	Internal structure					
		Geographic concentration					
Government and Culture	Government Support	Financial support systems					
		environmental regulations					
	Culture	Impact of national culture					
		Business climate					

Annexure 5: Information on Benefits After Implementing Quality Management Practices

Please tick information on Company's experience after implementing quality management practices (during the last 3 years)

S. No.	Company's Experience	Substantially Increased	Increased	No Change	Somewhat reduced	Substantially reduced
1.	Employee participation					
2.	Employee training					
3.	Employee morale					
4.	Competitive spirit					
5.	Human resource management					
6.	Manufacturing lead time					
7.	Energy consumption					
8.	Total quality cost					
9.	Warranty expenditure					
10.	Sales volume					
11.	Customer complaints					
12.	Company profit					
13.	Quality awards					
14.	Rework and scrap					
15.	Others					

Annexure 6: Information on Critical Success Factors for Adoption of Quality Management Practices

Please tick on critical success factors which may help MSMEs in adoption of Quality Management Practices on 1–5 scale with 1 = Not important, 2 = Somewhat important, 3 = Moderately important, 4 = Important, 5 = Very important

S. No	Factors	Sub criterions	1	2	3	4	5
1	Strategic factors	Top-management commitment					
		Quality culture					
		Quality systems					
		Quality awards					
		Continuous improvement and innovation					
		Benchmarking					
		Technical expertise					
		Marketing expertise					
		Financial access					
2	Tactical factors	Employee encouragement					
		Employee involvement					
		Training and education					
		Teamwork					
		Information and analysis					
		Supplier management					
		Communication					
3	Operational factors	Product and service design					
		Process management					
		Customer focus					
		Human resource management					
		Industrial relations					
		Employee rewards and incentives					

References

Nanda T, Singh TP (2009) An assessment of the technology innovation initiatives in the Indian small manufacturing industry. Int J Technol Policy Manag 9(2):173–207

Ooi K, Cheah W, Lin B et al (2012) TQM practices and knowledge sharing: An empirical study of Malaysia's manufacturing organizations. Asia Pac J Manag 29:59–78. https://doi.org/10.1007/s10490-009-9185-9

Prajogo DI (2005) The comparative analysis of TQM practices quality performance between manufacturing and services firms. Int J Serv Ind Manag 16(1):217–228

Sharma M, Kodali R (2008) TQM implementation elements for manufacturing excellence. TQM Mag 47(8):1281–1299

Subrahmanya MHB (2005) SMEs in India: will they be able to join global chains? The chartered accountant, sept, pp 421-427

Talib F, Rahman Z, Qureshi MN (2013) An empirical investigation of relationship between total quality management practices and quality performance in Indian service companies. Int J Qual Reliab Manag 30(3):280–318

Tari JJ (2005) Component of successful total quality management. TQM Mag 17(2):182–194

Tari JJ, Sabater V (2004) Quality tools and techniques: are they necessary for quality management? Int J Prod Econ 92(3):267–280

Tarí JJ, Molina JF, Castejón JL (2007) The relationship between quality management practices and their effects on quality outcomes. Eur J Oper Res 183(2):483–501

Zakuan NM, Yosf SM, Laosirihongthong T, Shaharoun AM (2010) Proposed relationship of TQM and organization performance using structural equation modeling. Total Qual Manag 21(2):185–203

Chapter 4
Need of Quality Management Practices

Highlights

Description of various reasons for introducing quality management practices in MSMEs based on 15 points related to quality consciousness, customer requirements, productivity improvement, etc.

4.1 Introduction

This chapter presents the information provided by the respondent organizations related to the prominent reasons for the introduction of quality management system in their organizations. It consists of the response rate, percentage of responses, type of industry involved and the reasons for introducing QMS in MSMEs.

4.2 Data Collection

Based upon the sample size, and having discussion with Industries Department, Government of Himachal Pradesh survey was carried out in four main industrial clusters of the state, i.e.,***Kangra, Solan, Sirmour*** and ***Una***. The survey covered 379 MSME units in the above mentioned clusters. Out of the total 379 questionnaires administered through postal, e-mail and personal visits, 286 were found appropriate, thus yielding a response rate of 76.26%, which is the sufficient number for further analysis. It was ensured that the questionnaire is to be completed by the person responsible for managing quality aspects in the company. They were mainly CEOs or top management representatives, having firsthand knowledge of quality implementation in their companies. The data collection was done in four phases these are:

R .K. Sharma, *Quality Management Practices in MSME Sectors*,
Springer Tracts in Mechanical Engineering,
https://doi.org/10.1007/978-981-15-9512-7_4

Table 4.1 Respondents profile

	Work profile	Percent (%)	Working experience (years)	Education profile	Percent (%)
Respondent profile	Quality manager	30.54	10–15	PhD degree	5
	Plant manager	32.24	20–40	Postgraduate	24
	Production Manager	19.15	15–20	graduate	43
	Engineering Manager	18.07	10–15	Under-graduate	28

i. **Phase 1**
ii. **Phase 2**
iii. **Phase 3**
iv. **Phase 4**

It has taken around a complete 1 year to collect data from these industrial clusters because of complex geographical conditions of the state.

4.3 Respondents Profile

Among the respondents, 30.54% were quality managers, 32.24% were plant managers; 19.15% production managers and remaining 18.07% were management representatives. The more details of respondents, working experience and educational profile are presented in Table 4.1.

4.4 Data Analysis

4.4.1 Response Profile

Data collected with respect to the first aspect, i.e., general information of company type and year of establishment revealed (Table 4.2) that a large proportion, i.e., 65% of the organizations are Private Ltd., followed by 34%, as partnership and 1% as joint venture organizations. As per the Table 4.3 (page 47), the maximum number of responses is from pharmaceutical sector 80 units (28%), followed by mechanical and electrical and electronics sector 60 units (21%) each, food and allied 34 units (12%), paper and paper products 23 units (8%), and textile 9 units (3%), etc. The maximum numbers of units were established after year 2003, which shows that most of the MSMEs in the region are new and progressing.

Table 4.2 Type of organization surveyed

Company type	Number of respondents	Percent (%)
Private Ltd	184	65
Government	1	0
Joint Venture	4	1
Partnership	97	34
Total	**286**	**100**

Table 4.3 Classification of respondents industries(Investment in equipments and machinery)

Category of firm	No. of respondents	Percent (%)
Micro Scale Unit (<25 Lakhs)	44	15
Small Scale Unit (<5 Crores)	190	67
Medium Scale Unit (<10 Crores)	52	18
Total	**286**	**100**

Further, the survey found that majority 190 (67%) of industries were small scale followed by medium (18%) and micro (15%), as shown in Table 4.3 respectively.

Table 4.4 shows the classification of industries as per the norms of pollution control board. The survey found that majority of industries, i.e., 159 (56%) are under green category, followed by 117 (41%) under orange and remaining only 3% are under red category. The detail with respect to definitions of green, orange and red categorization is provided in Annexure-III.

The next question was on awareness about quality aspects. Survey result as shown in Table 4.6 found that 99% organizations were aware about quality (Table 4.5).

Table 4.4 Categorization as per pollution control board

Categorization	Number of respondents	Percent (%)
Green	159	56
Orange	117	41
Red	9	3
Not replied	1	0
Total	**286**	**100**

Table 4.5 Awareness of quality among MSMEs

Awareness about quality	Number of respondents	Percent (%)
Yes	285	99
No	1	1
Total	**286**	**100**

Table 4.6 Firms with ISO certifications

ISO certified	No. of respondents	Percent (%)
Yes	200	70
No	86	30
Total	**286**	**100**

Though about 100% firms were aware about quality but still 30% firms were not ISO certified. Total out of 284 firms 200 ticked on "Yes" whereas 86 on "No" when they were asked about ISO certification.

4.4.2 Need of Quality Management Practices

This section of questionnaire deals with 15 questions, which were designed to get information on need of quality management practices and quality standard certification in an organization. The respondents were asked to tick on boxes callipered with five-point Likert scale, i.e., 1 = Not important, 2 = somewhat important, 3 = moderately important, 4 = important, 5 = Very important. Table 4.6, 4.7 shows the summary of results obtained through responses with respect to fifteen points.

4.4.2.1 Quality Consciousness Among Employees

With respect to the first point, i.e., "to increase the quality consciousness among employees," it was found that 60% MSMEs in the region feel that it is important and 33% feel that it is very important as shown in Fig. 4.1 respectively.

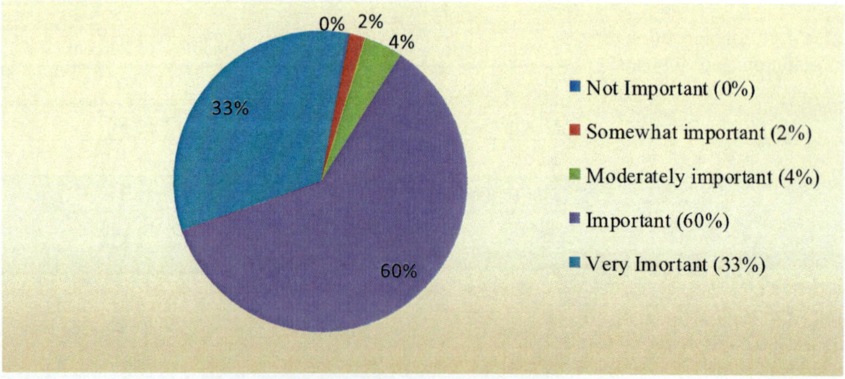

Fig. 4.1 To increase quality consciousness among employees

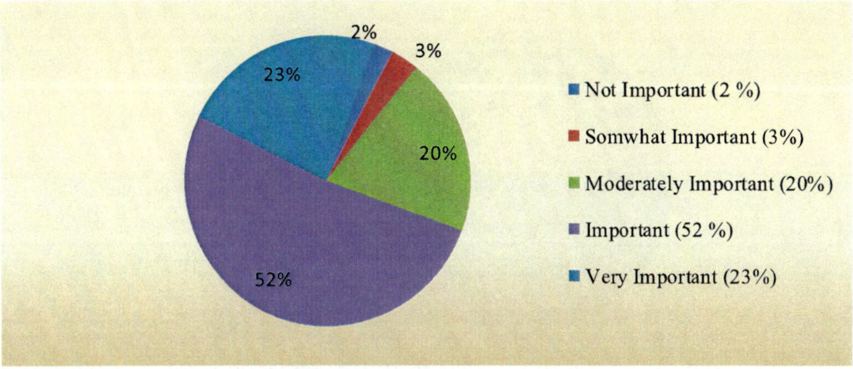

Fig. 4.2 Role of formalized system for documentation

4.4.2.2 Adoption of Formalized System for Documentation

With regard to second point "to adopt formalized system that documents the structure, responsibility and procedures for effective quality management practices," it was found that 52% respondent feels it important while 23% considered it very important and 20% moderately important as shown in Fig. 4.2.

4.4.2.3 Competitive and Successful Business Operations

With respect to the third point "to be competitive and successful in business operations both in domestic and foreign market," it was found that 36% of industries feel it very important where as 52% feels it important for implementing quality practices for improving their competitive positions in both domestic and foreign market as shown in Fig. 4.3 respectively.

4.4.2.4 Meet Customer Requirements and Ensure Consistency

The result pertaining to fourth point, i.e., "to meet customer requirements and ensure consistency in product and service delivery," as shown in Fig. 4.4, it was found that 40% firms feel it important, 51% firms feel it very important and 8% feel it moderately important to introduce quality management system, with total PPS score 87.75%. Thus, we can say that overall motive of introducing quality management practices is to satisfy customer's needs by providing better products and services.

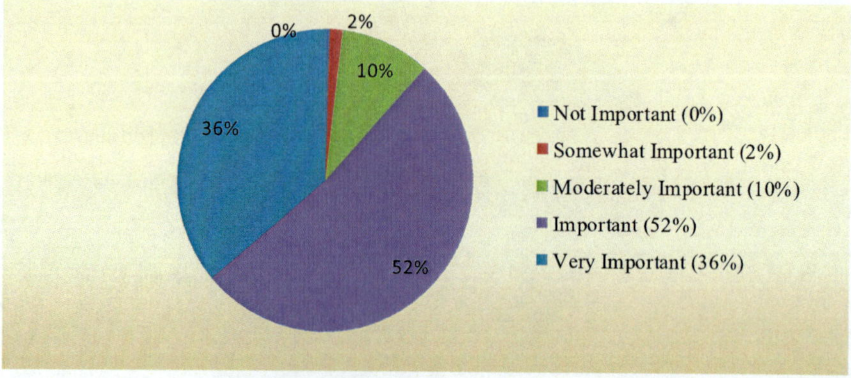

Fig. 4.3 To be competitive at domestic and foreign market

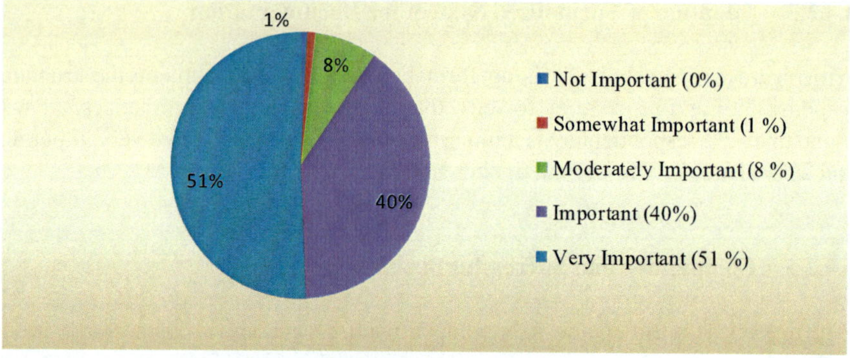

Fig. 4.4 To meet customer requirement through consistency in product and service

4.4.2.5 Promoting Lean Manufacturing

With regard to fifth question "to identify inefficient processes and promote lean manufacturing" only 17% of firms feel it very important and 41% important, with moderate PPS value, i.e., 73.1. This is because the concept of lean manufacturing in new and just at beginning stage in Indian MSMEs. The pie diagram is shown in Fig. 4.5.

4.4.2.6 Withstand Pressure from Existing Supplier(s)

With regard to sixth point "to withstand pressure from existing suppliers," the results presented in Fig. 4.6 show that maximum 34% firms feel it as moderately important, 12% firms feel it very important even some firms 6% feels it not an important reason to introduce a quality management system as indicated by low PPS score, i.e., 63.03%.

Table 4.7 Reasons for introducing QMS and gaining quality standard certification in a company

S. No	Reasons	1	2	3	4	5	PPS
1	To increase quality consciousness among the employees	1	5	12	172	94	84.86
2	To adopt a formalized system that documents the structure, responsibilities and procedures for effective quality management	6	9	56	147	66	78.17
3	To be competitive and successful in business operations both in domestic and foreign market	1	4	28	148	103	84.51
4	To meet customer requirements and ensure consistency in products and service delivered	2	3	22	113	144	87.75
5	To identify inefficient processes and promote lean manufacturing	5	19	92	117	48	73.10
6	To withstand pressure from existing supplier(s)	16	65	97	72	34	63.03
7	To reduce potential audit expenditure both internal as well as external	32	84	70	65	33	58.80
8	To reduce warranty and replacement expenditures	71	83	46	54	29	52.01
9	To reduce the quality costs, i.e., prevention, appraisal and failure costs	7	49	123	77	28	64.93
10	To upgrade skills/training of employees	2	4	17	180	80	83.46
11	To increase productivity	1	1	5	75	200	93.48
12	To increase profitability	1	4	8	89	180	91.42
13	To develop a nonpolluting environment-friendly manufacturing process with minimal waste	3	10	55	161	53	77.80
14	To reduce potential energy expenditure	2	6	60	168	47	77.81
15	To sustain long-term growth, dynamism and employment opportunities	1	1	10	94	178	91.48

Where: 1 = Not important, 2 = Somewhat important, 3 = Moderately important, 4 = Important, 5 = Very important; PPS = Percent Point Score.

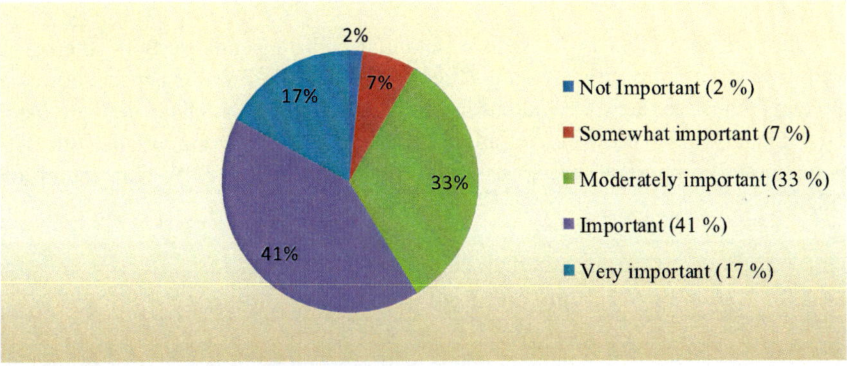

Fig. 4.5 To promote lean manufacturing

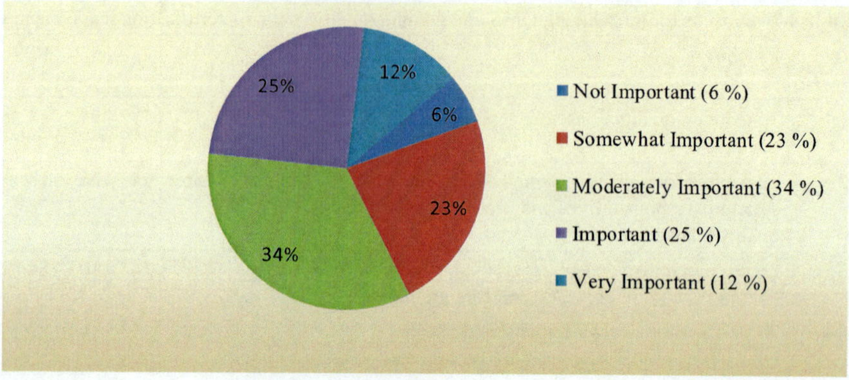

Fig. 4.6 To withstand pressure from supplier pressure

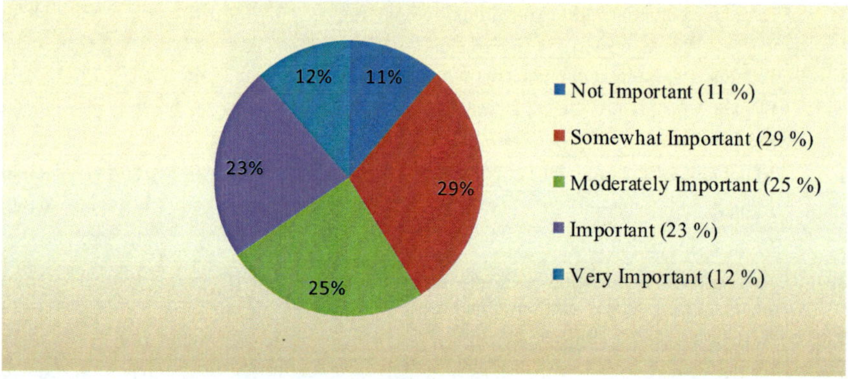

Fig. 4.7 To reduce internal/external audit expenditure

4.4.2.7 Reducing Potential Audit Expenditure

The next seventh point is "to reduce potential audit expenditure both internal as well as external" the survey results (Fig. 4.7) with regard to this aspect show that the companies that are aware about importance of quality practices they are using it for their own satisfaction. As only 12% firms feel it very important with 25% as moderately important and 23% as important. The overall PPS score found low (58.80).

4.4.2.8 Reducing Warranty and Replacement Expenditure

With regard to the eighth point "to reduce warranty and replacement expenditure the overall PPS point were found quite low, i.e., 52.01% only, it comprises of 10% very

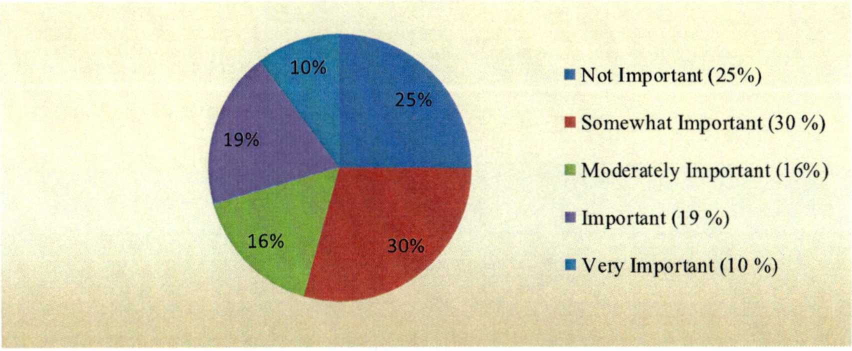

Fig. 4.8 To reduce warranty and replacement expenditure

important, 19% important, 16% moderately important, 30% somewhat important and 25% not important observations as revealed from respondents (Fig. 4.8). In nutshell, we can say that introducing a quality management system helps the firms in reducing warranty and replacement expenditure.

4.4.2.9 Reducing the Quality Costs

With respect to the ninth point, i.e., to reduce the quality costs namely, prevention, appraisal, and failure costs. The result with respect to this aspect is presented in Fig. 4.9 with 43% firms found it moderately important, whereas 27% feel it important. The overall PPS score from this construct was found 64.93%, which shows huge scope of improvements and that is possible only when companies go for quality cost accounting system.

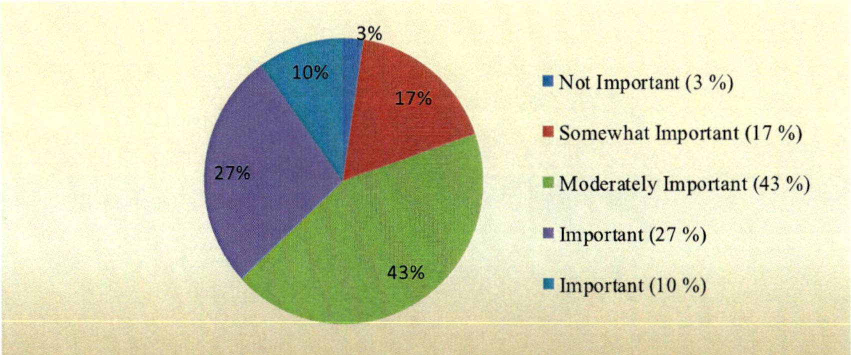

Fig. 4.9 To reduce quality costs

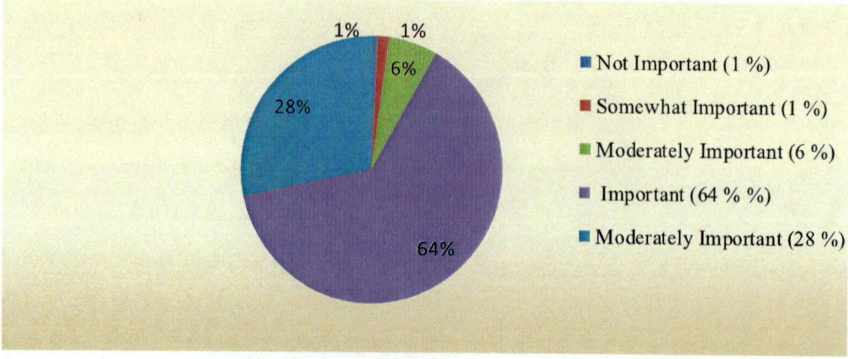

Fig. 4.10 To upgrade skills/training of employees

4.4.2.10 Upgrading the Skills and Training of Employees

With respect to 10th point, i.e., to upgrade skills/training of employees, the result shows that 28% firms feel it is very important and 64% important, with the overall PPS score comes to 83.46%. The result with respect to this point is shown in Fig. 4.10.

4.4.2.11 Increasing Productivity

With regard to 11th point "to increase productivity" the result found that 71% firms feel it very important, 27% important, only 1% ticked on not important and 2% feel moderately important, which comprises high overall PPS score 93.48%. This high PPS score shows that introducing quality management system plays vital role in increasing productivity. The results are presented in Fig. 4.11.

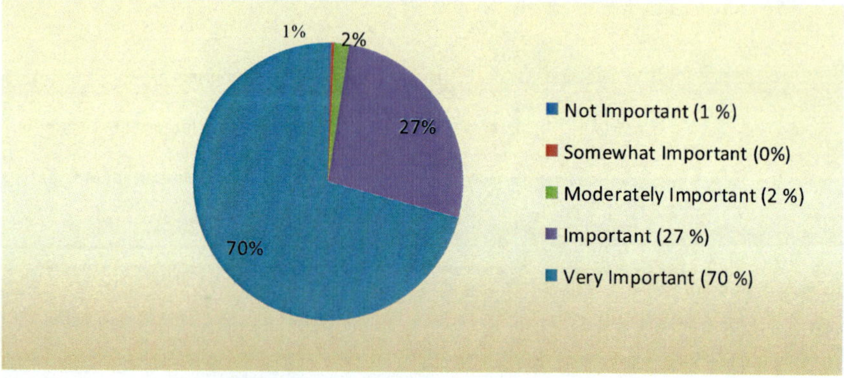

Fig. 4.11 To increase productivity

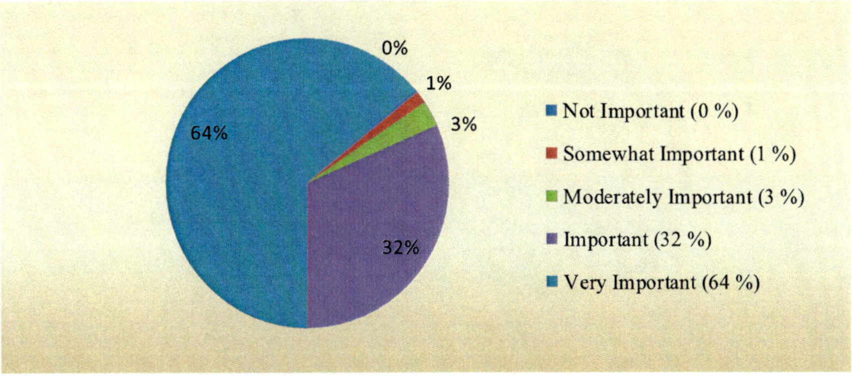

Fig. 4.12 To increase profitability

4.4.2.12 Increasing Profitability

With regard to 12th point to increase profitability results are shown in Fig. 4.12. It can be seen that 64% MSMEs feel it very important whereas 32% important, with overall PPS score 91.42%. From this, it can be concluded that productivity is directly related to profitability, so with quality performance.

4.4.2.13 Developing a Nonpolluting Environment-Friendly Manufacturing Process

With regard to point number 13, i.e., "to develop a non-polluting environment-friendly manufacturing process with minimal waste" the results shown in Fig. 4.13 revealed that 57% firms accepted that introducing a QM system plays an important

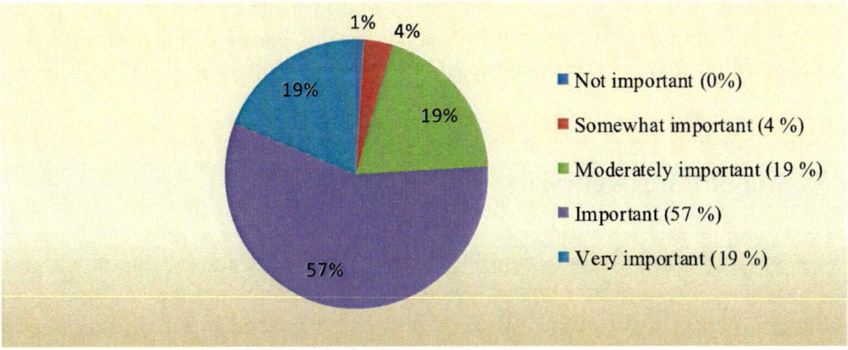

Fig. 4.13 To minimize waste

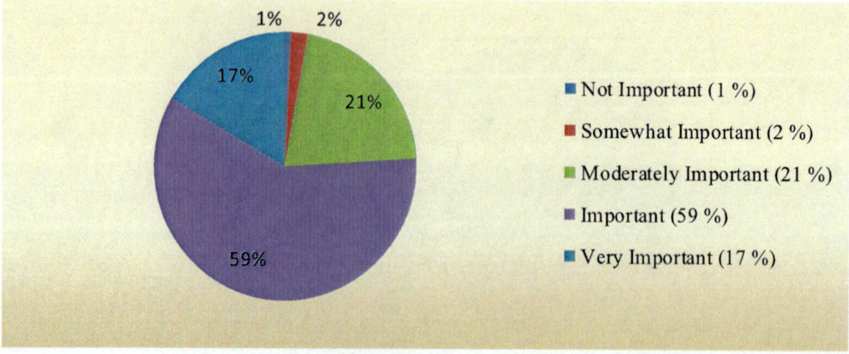

Fig. 4.14 To reduce potential energy expenditure

role in creating nonpolluting environment, with 19% important and same percentage for moderately important. The overall PPS score was found to be 77.80%.

4.4.2.14 Reducing Potential Energy Expenditure

With respect to the 14th point "to reduce potential energy expenditure" the results are presented in Fig. 4.14. It can be seen that 17% firms feel it very important, 59% important with 21% found it moderately important. The overall PPS score for this construct obtained is 77.81%.

4.4.2.15 Sustaining Long-Term Growth, Dynamism and Employment Opportunities

With respect to 15th point, i.e., "to sustain long-term growth, dynamism and employment opportunities" 63% firms feel it very important and 33% important and 4% moderately important. The overall PPS score for this construct is found high, i.e., 91.48%, which shows the importance of quality management (Fig. 4.15).

4.5 Concluding Remarks

The chapter presents the details to the type of the organizations, classification of respondents companies and their categorization into green, orange and red as per pollution control board norms. This chapter has described in detail, the various reasons as to why MSMEs wish to introduce quality management system? The discussion is based on 15 points. As shown in Fig. 4.16 that the prominent reasons for introduction of quality management system are to increase productivity with high

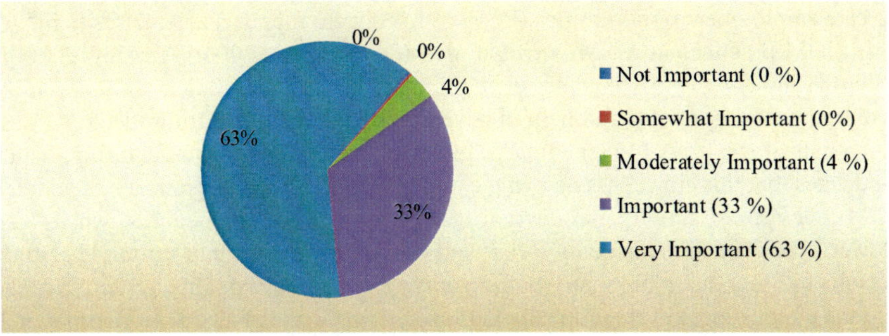

Fig. 4.15 Long term growth, dynamism and employement

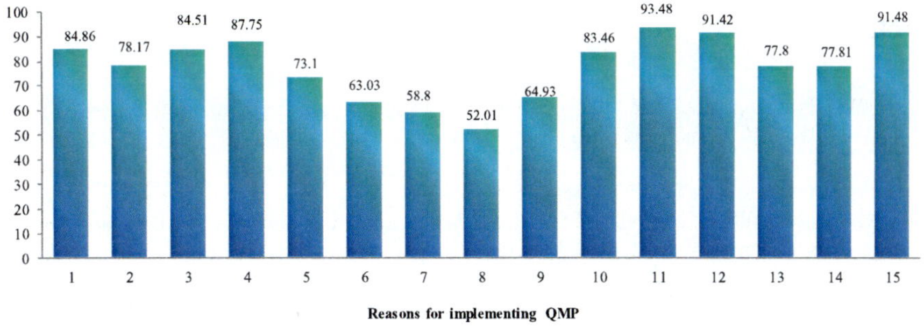

1. To increase quality consciousness among the employees, 2. To adopt a formalized system that documents the structure, responsibilities and procedures for effective quality management, 3. To be competitive and fruitful in business operations both domestic and foreign market, 4. To meet customer requirements and ensure consistency in products and service delivered, 5. To identify inefficient processes and promote lean manufacturing, 6. To withstand pressure from existing supplier(s), 7. To reduce potential audit expenditure both internal as well as external, 8. To reduce warranty and replacement expenditures, 9. To reduce the quality costs, i.e., prevention, appraisal and failure costs, 10. To upgrade skills/training of employees, 11. To increase productivity, 12. To increase profitability, 13. To develop a nonpolluting environment-friendly production process with minimal waste, 14. To reduce possible energy expenditure, 15. To sustain long-term growth, dynamism and employment opportunities.

Fig. 4.16 PPS score for implementing quality management practices

PPS score (93.48) followed by sustaining long-term growth dynamism and employment opportunities with PPS score (91.48) and to increase profitability (Lewis et al 2006). The majority of organization implemented QM practices to impart the quality consciousness among their employees; as they considered it essential to enhance the performance at the workplace. The study results are in line with Karia and Asaari (2006); Ooi et al. (2012). Most of the organization (68%) started QM practices to attain the benefits from employees training provided during QM implementation procedure. All these findings are supportive of literature findings (Sadikoglu and Zehir 2010; Kharub and Sharma 2018, 2020). Apart from the meeting customer requirements and ensuring consistency in product delivery (PPS = 87.75), developing nonpolluting environment (77.80) in order to sustain long-term dynamism and create employment opportunities (PPS = 91.48) were found as key factors to implement QM practices in MSMEs (Sit et al 2009; Wang et al 2016; Terziovski 2006; Sharma and Kharub 2015; Dubey et al 2015). QM techniques reduce waste, promote continuously improving and help in developing lean manufacturing ideology. Less than half of the firms (46%) employed QM to promote lean manufacturing, which indicates that this concept is new in the region. Further, some organizations adopted QM practices to simplify production tasks and workflow procedures, which lead toward the reduction in quality cost and reduction in warranty costs. The study results indicate that more than a half organization adopted a quality system to speed up the effectiveness of their documentation system (Brikc et al. 2013; Matthias et al 2013).

References

Lewis WG, Pun KF, Lalla TRM (2006) Empirical investigation of the hard and soft criteria of TQM in ISO9001 certified small and medium-sized enterprises. Int J Qual Reliab Manag 23(8):964–985

Karia N, Asaari MHAH (2006) The effect of total quality management practices on employee's work-related attitudes. The TQM Mag 18(1):30–43

Ooi KB, Cheah WC, Lin B, The PL (2012) Total quality management practices and knowledge sharing: an empirical study of Malaysia's manufacturing organizations. Asia Pac J Manag 29(1):62–77

Sadikoglu E, Zehir C (2010) Investigating the effects of innovation and employee performance on the relationship between TQM practices and firm performance: an empirical study of Turkish firms. Int J Prod Econ 127(1):13–26

Sit WY, Ooi KB, Lin B, Chong AYL (2009) TQM and customer satisfaction in Malaysia's service sector. Ind Manag Data Syst 109(7):957–975

Wang Q, Zhao X, Voss C (2016) Customer orientation and innovation: A comparative study of manufacturing and service firms. Int J Prod Econ 171:221–230

Terziovski M (2006) Quality management practices and their relationship with customer satisfaction and productivity important. Manag Res News 29(7):414–424

Sharma RK, Kharub M (2015) Qualitative and quantitative evaluation of barriers hindering the growth of MSMEs. Int J Bus Excell 8(6):724–747

Dubey R, Gunasekaran A, Ali SS (2015) Exploring the relationship, operational practices institutional pressure and environmental performance: A framework for green supply chain. Int J Prod Econ 160(1):120–132

Brkic VK, Djurduja T, Dondur N, Klarin MM, Tomic B (2013) An empirical examination of the impact of quality tools application on business performance: Evidence from Serbia. Total Qual Manag 24(5):607–618

Kharub M, Sharma RK (2018) Quantifying the relationship between latent variables after successful implementation of QM practices in MSMEs. Int J Q Reliab Manag 35(4): 875–896. https://doi.org/10.1108/IJQRM-12-2016-0221.ISSN: 0265-671X

Kharub M, Sharma RK (2020) An integrated structural model of QMPs, QMS and firm's performance for competitive positioning in MSMEs, Total Q Manag Bus Excell (Online published) 2020. ISSN: 1478-3363

Matthias T, Moacir GF, Mark S, Lawrence DF (2013) Competitive priorities of small manufacturers in Brazil. Ind Manag Data Syst 113(6):856–874

Chapter 5
QMS Implementation Stages

Highlights

- Various sectors of MSME operation.
- Year-wise Adoption of Quality Practices/Programs.
- Stages in company where QM practices are applied.

5.1 Introduction

The various sectors of operation of MSME include, Chemical and allied Food and allied, Mechanical, Electrical and Electronics, Textile, Paper and Paper Products, etc. In this chapter, the year-wise adoption of quality practices, stages in company where quality management practices are applied and nature of quality management tools being applied has been discussed. Data were collected from four industrial clusters (as mentioned in Chap. 4) are presented with the help of various figures.

5.2 Sector of Operation

Figure 5.1 presents the sector-wise operation of various firms in the four main industrial clusters of the state, i.e., ***Kangra***, ***Solan***, ***Sirmour***, and ***Una***. As evident from Fig. 5.1 that 28% of them are chemical and allied (Pharmaceutical), 21% of them are both electrical and electronics and mechanical, 12% of them are food and allied, 8% of them are paper and paper products, 3% textile, 2% each glass and leather and 1%

R .K. Sharma, *Quality Management Practices in MSME Sectors*,
Springer Tracts in Mechanical Engineering,
https://doi.org/10.1007/978-981-15-9512-7_5

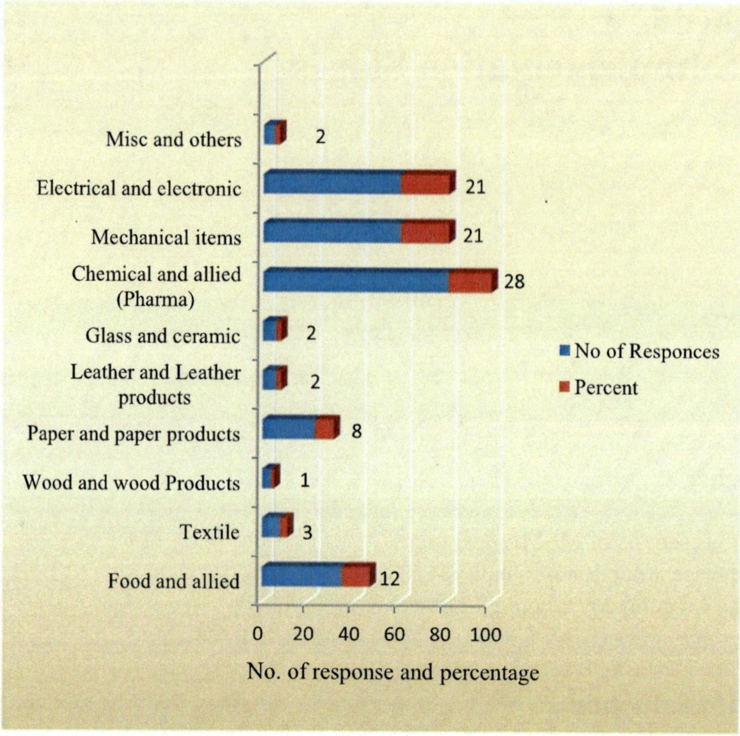

Fig. 5.1 Sector wise operation of firms

wood and wood products and 2% miscellaneous and others firms. This shows that pharmaceutical companies have more density as compared with other sectors. This is the reason why this region is called hub of pharmaceutical companies.

5.3 Year-wise Adoption of Quality Practices/Programs

With regard to year-wise adoption of quality practices, the collected statistic shows that 38% of firms have adopted quality practices during the year 2006–2010, followed by 25% in year 2011 onward. Very few firms adopted quality programs during 1996–2000 and 2001–2005, respectively, as shown in Fig. 5.2. The results show that with increase in awareness regarding quality among the MSMEs, the adoption of quality practices has increased up to 73% in cumulative.

The results about the effect of time since the adoption of ISO-9000 certification on QM practices implemented are shown in Fig. 5.3. Further, the influence of time on the adoption of ISO and firm size on QM practices is investigated. It was found

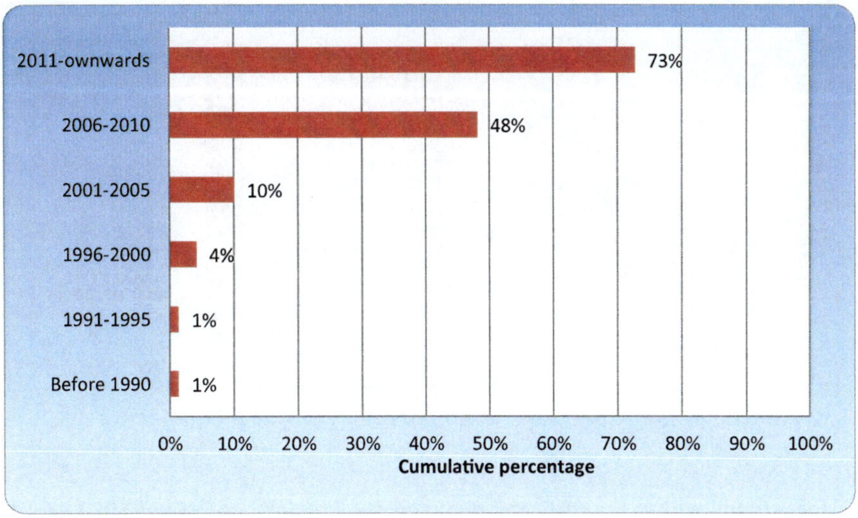

Fig. 5.2 Year-wise adoption of quality management practices

Fig. 5.3 Influence of time since adoption of ISO and firm size on QM practices

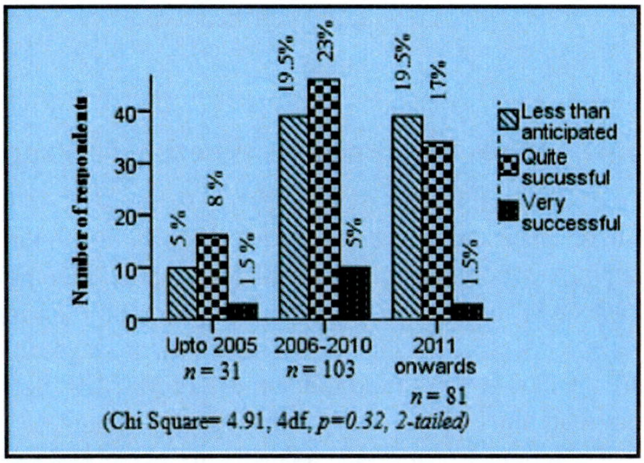

that time since the adoption of ISO certificate was not statistically significant with QM practice implementation ($\chi^2 = 4.91$, 4df, $p = 0.32$), so, the study rejects the possibility that the time duration after the adoption of ISO-9000 certificate does effect on the implementation of QM practices.

The representatives of companies were asked regarding awareness of quality management and whether their firm is ISO certified or not. A total of 285 valid responses were received on this aspect. The detailed results are presented in Fig. 5.4. Though all firms were aware of quality management still, 25% companies were not ISO certified. Overall, 75.4% (215) companies were found ISO certified, which comprises of 11.2% micro, 36.8% small and 27.4% medium-size companies. Of the medium size firms, 88% companies were found ISO certified, while 75% of small size

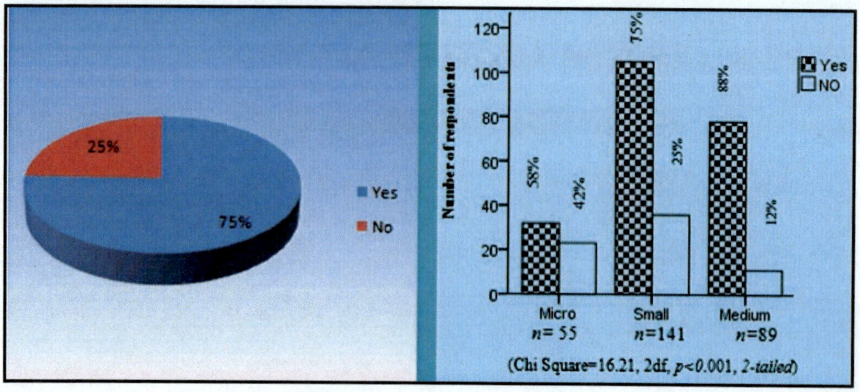

Fig. 5.4 Firms with ISO certification

business found with ISO certificates. Only, 58% microunits were considered having ISO certification. Study results found meaningful ($\chi^2 = 16.21$, 2df, $p < 0.001$) association between firm size and ISO certification. Based on these results, we can say that compared with microunits, medium and small size companies have significant devotion in adopting quality certification.

5.4 Stages in Company Where QM Practices Applied

There are usually five stages in which quality management practices have been applied by companies in MSME sectors, i.e., incoming raw material/vendors place inspection, in-process inspection, finished product inspection, packaging and after sales service. In the study, with regard to the stages in company where QM practices are applied, it was found that almost all MSMEs, i.e., 98% are inspecting incoming raw material, 87% are using them in process, and 92% are using them at finished stage. It was found that only 4% MSMEs are using these practices after sales. Figure 5.5 shows the details with respect to stages where these quality management practices are being applied by the firms. The representatives of companies were asked regarding their zeal (High, Medium, Low) of acquiring quality awards. A total of 212 replies were received. The detailed results are presented in Fig. 5.6. The quality award gives special recognition to the company, which strives for excellence by generating interest in quality programs and meets challenges of implementing quality systems. Study results show just 24% industries make tremendous efforts to pursue these awards whereas 48% are just starting spreading awareness among all departments. The remaining 28% industries are found unaware about quality awards. Hence, they are not making any effort for it. The most of the industries were considered familiar with Rajiv Gandhi National Quality Award (RGNQA) as discussed by one of the RGNQA holders, i.e., Elin Appliances Private Limited, Solan, Himachal Pradesh.

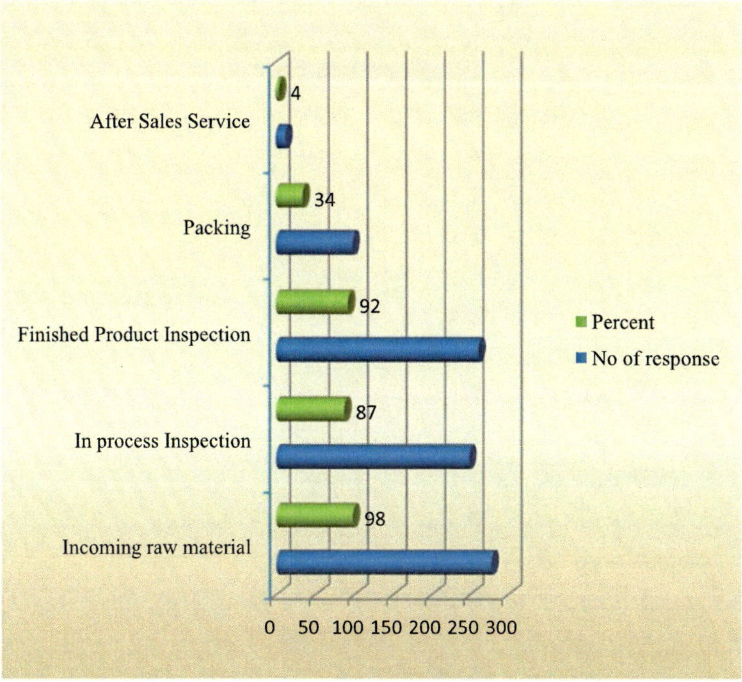

Fig. 5.5 Stage where QM practices are being used

Fig. 5.6 Enthusiasm of acquiring quality awards

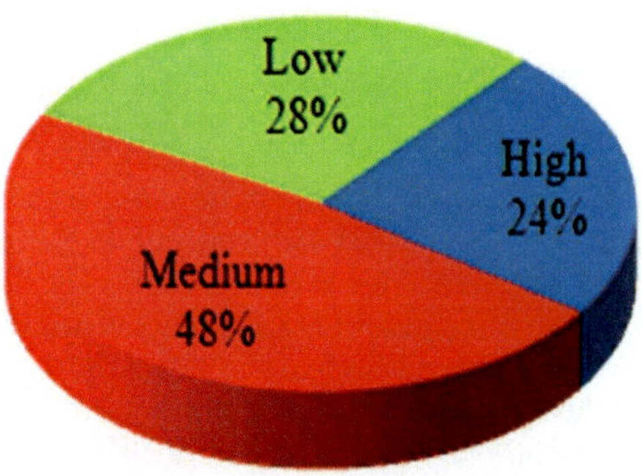

5.5 Nature of Quality Management Tools Being Applied

Further, with respect to nature of quality management tools applied, the results are presented in Fig. 5.7. As shown in Fig. 5.7, it was found that maximum (81%) MSMEs are using check sheets, 71% MSMEs have implemented 5 S, 47% MSMEs are using control charts, 43% MSMEs use cause and effect diagrams, 32% MSMEs use customer relationship management, 15% MSMEs use enterprises resource planning, 8% MSMEs practice each six sigma and process flow diagram, 4% MSMEs use

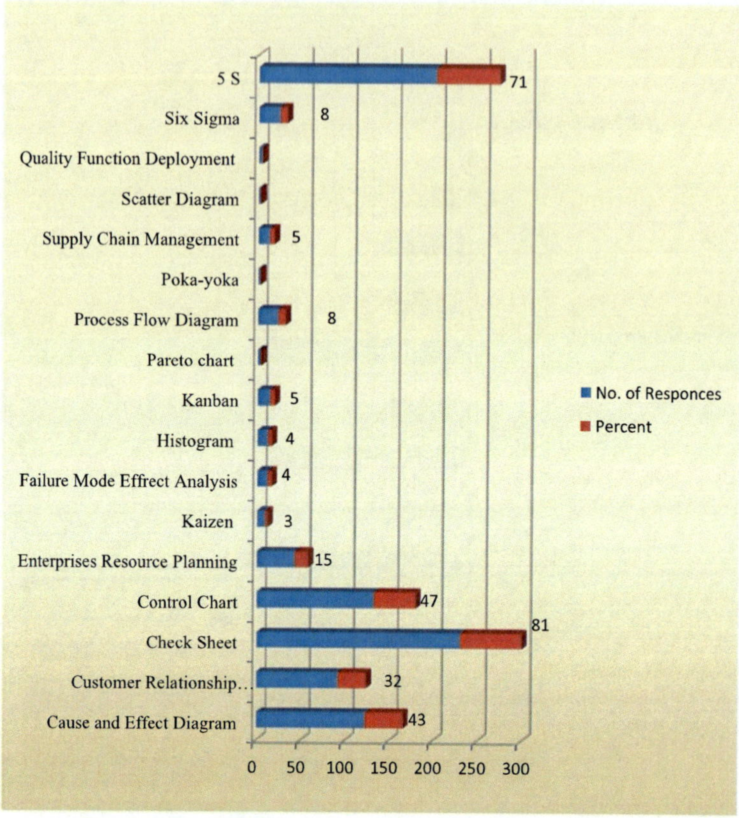

Fig. 5.7 Nature of QM tools being used by MSMEs

each histogram and failure mode and effect analysis, 5% MSMES are using Kanban and only 3% MSMEs are familiar with Kaizen concept. The findings indicate the weakness of MSMEs in lack of implementation of tools and techniques for quality improvement, mainly regarding advanced tools such as Six Sigma, histogram, failure mode and effect analysis, etc. Table 5.1 presents the firm wise details with respect to application of QM tools. Companies in sectors, i.e., pharmaceutical, Electrical and Electronics and Mechanical use most of the quality tools as compared with other sectors.

5.6 Concluding Remarks

The results indicate that with an increase in awareness regarding quality among the MSMEs, the adoption of quality practices has grown up to 73% in cumulative. 75.4% companies were found ISO certified, which comprises of 11.2% micro, 36.8%

Table 5.1 Firms-wise details of QM tools

Industrial sectors ⇨ / Quality tools ⇩	Total	Electrical and Electronics	Mechanical	Pharmaceutical	Glass and ceramics	Leather and leather products	Paper and paper products	Wood and wood products	Textile	Food and allied
Cause and effect diagram	**123**	32	45	31	4	0	5	0	4	2
Customer relationship management	**92**	30	33	19	0	0	4	0	3	3
Check sheet	**231**	49	55	68	3	2	19	4	6	25
Control chart	**133**	34	46	25	0	0	12	0	2	14
Enterprise resource planning	**43**	12	14	12	0	0	5	0	0	0
Kaizen	**9**	2	4	3	0	0	0	0	0	0
Failure mode and effect analysis	**12**	2	8	2	0	0	0	0	0	0
Histogram	**12**	3	5	4	0	0	0	0	0	0
Kanban	**14**	0	11	3	0	0	0	0	0	0
Pareto chart	**3**	0	3	0	0	0	0	0	0	0
Process flow diagram	**23**	5	12	4	0	0	0	0	0	2
Poka-Yoke	**2**	0	2	0	0	0	0	0	0	0
Supply chain management	**13**	2	3	5	0	0	1	0	1	1
Scatter siagram	**2**	0	2	0	0	0	0	0	0	0
Quality function	**4**	0	4	0	0	0	0	0	0	0
Six-Sigma	**24**	11	13	0	0	0	0	0	0	0
5 S	**201**	45	48	66	3	2	12	0	4	21

small and 27.4% medium size companies. Three critical stages of inspection were found as (i) inspection of incoming raw material, (ii) inspection during the process and (iii) inspection of the finished component, whereas, only 4% MSMEs are using these practices after the sale. It was found that maximum (81%) organizations are using check sheets, while 5S (sort, set in order, shine, standardize, and sustain) is noted second most (71%) acceptable quality management tools among MSMEs. The findings indicate the weakness of MSMEs in lack of implementation of tools and techniques for quality improvement, mainly regarding the advanced tools such as Six Sigma, histogram, failure mode and effect analysis, etc.

Chapter 6
Information on Competitiveness

Highlights

Extensively known Porter diamond framework has been used in literature to measure and analyze the competitive advantage by business units using determinants with casual and proxy variables related to factor conditions, demand conditions, firm's strategy and structure, and the presence of related and supported industries and one external determinant government and culture.

6.1 Introduction

This section presents the results on determinants of competitiveness among MSME sectors based upon Diamond Porter's Model. The four key determinants of the diamond model are (i) factor conditions, (ii) demand conditions, (iii) firm's strategy and structure, and (iv) presence of related and supported industries and one external determinant is government and culture. These five determinants were measured with the help of proxy variables and the responses from the questionnaire survey in selected industrial clusters, as obtained from MSME respondents are computed and are tabulated in Table 6.1, respectively.

6.2 Factor Conditions

The factor conditions have been divided into two casual variables, i.e., basic factors and advance factors. Proxy variables used to measure the basic factors are, national resource, physical resource, unskilled labor, highly educated person and proxy

R .K. Sharma, *Quality Management Practices in MSME Sectors*,
Springer Tracts in Mechanical Engineering,
https://doi.org/10.1007/978-981-15-9512-7_6

Table 6.1 Determinants with casual and proxy variables to measure competitive advantage; where: 1—Strongly disadvantage 2—disadvantage, 3—Neither advantage-Nor disadvantage, 4—advantage and 5—strongly advantage; PPS: Percent Point Score

Determinants	Casual variables	Proxy variables	1	2	3	4	5	PPS
Factor conditions	Basic factors	Natural resources	1	5	39	179	62	80.70
		Physical resources	2	7	61	175	40	77.12
		Unskilled labor	10	26	53	98	98	77.40
		Highly educated personnel	2	7	38	138	100	82.95
	Advanced factors	Production and process technologies	1	2	25	140	118	86.01
		IPR and patents	97	102	46	21	20	43.57
		Communication infrastructure	1	8	55	194	28	76.78
Demand conditions	Market volume	Market size	2	0	20	185	78	83.65
		Pattern of growth	1	0	29	202	54	81.54
	Sophistication	Distribution channels	3	7	46	180	46	78.37
		New investment in region	1	2	48	185	49	79.58
Related and supporting industries	Related companies	Technology upgrading	2	2	17	95	170	90.00
		Information flow	0	3	45	190	47	79.72
		Shared technology development	0	5	32	156	91	83.45
	Support	R&D investments	4	21	80	88	91	76.97
		Suppliers and distribution channels	0	8	50	184	44	78.46
		Marketing	0	7	32	173	73	81.89
Firm structure strategy, and rivalry	Rivalry	Boosted innovation	0	5	34	195	52	80.56
	Structure/strategy	Internal structure	0	5	28	204	49	80.77
		Geographic concentration	1	6	32	140	105	84.08
Government and culture	Government support	Financial support systems	2	3	27	120	134	86.64

(continued)

Table 6.1 (continued)

Determinants	Casual variables	Proxy variables	1	2	3	4	5	PPS
		Environmental regulations	1	1	53	189	40	78.73
	Culture	Impact of national culture	1	2	41	213	29	78.67
		Business climate	1	2	27	182	73	82.74

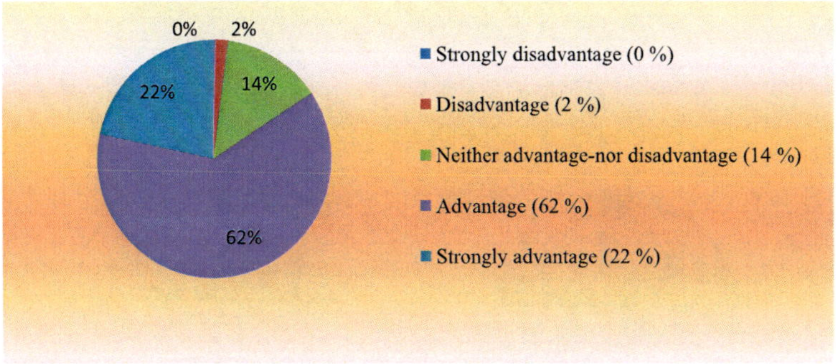

Fig. 6.1 Competitive advantage through natural resources

variables used to measure the basic advance factors are production and process technology, IPR and patents, communication infrastructure, respectively.

6.2.1 *Presence of Natural Resource*

With regard to the first variable, i.e., "presence of natural resource," 62% of the MSMEs feel that these are advantageous factors, where as 22% describe it as strongly advantageous factors for competitive positioning. Overall item's PPS score comes out to be 80.70%, which indicates the importance natural resources (Fig. 6.1).

6.2.2 *Role of Physical Resources*

With regard to next proxy variable, i.e., "role of physical resources" the result shows that 61% firms agreed that available physical resources are advantageous, and 14% strongly advantageous, for attaining competitive positioning as shown in Fig. 6.1. The overall PPS score for this construct was found to be 77.12, respectively (Fig. 6.2).

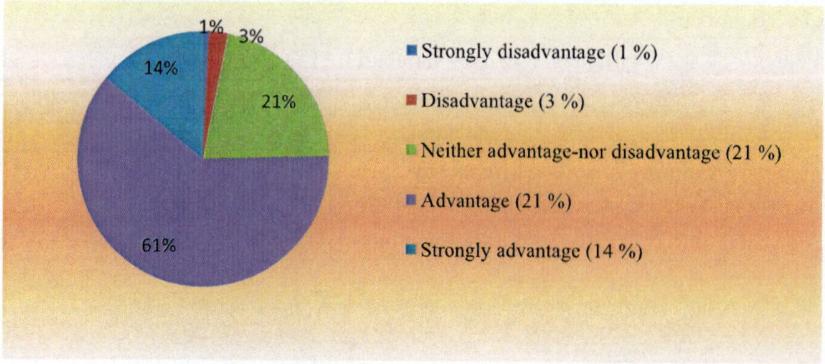

Fig. 6.2 Role of physical resources

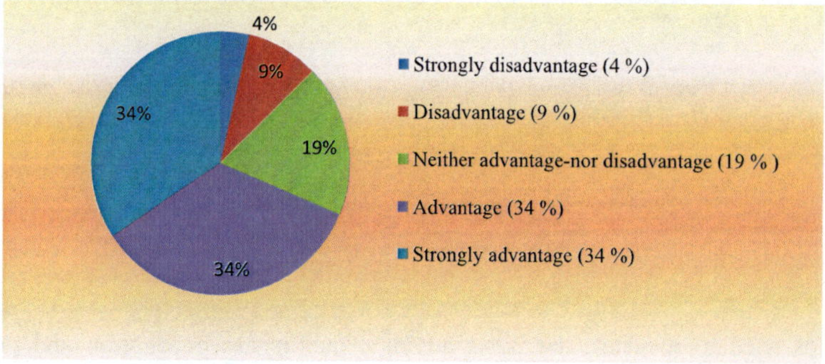

Fig. 6.3 Role of unskilled labour

6.2.3 Role of Unskilled Labor

With respect to next item, i.e., "Unskilled labor," 34% MSMEs feel it as advantageous and same percentage found it strongly advantageous. This shows that the presence of unskilled labor helps in carrying out their business operations effectively. The overall PPS score for this item was found 77.40. Figure 6.3 presents the details.

6.2.4 Role Highly Educated Personnel

With respect to the role of proxy variable, i.e., "highly educated personnel" in attaining competitive positioning in MSMEs, results are presented in Fig. 6.4. It was found that 35% firms feel it strongly advantageous and 48% advantageous. The overall PPS score for this item was found 82.95% which is quite high.

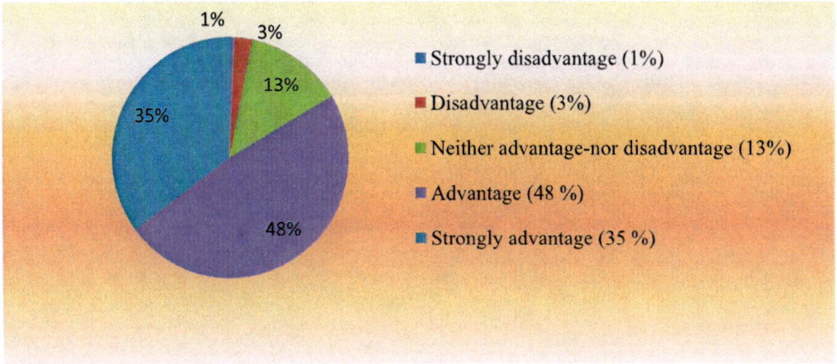

Fig. 6.4 Role of highly educated personnel

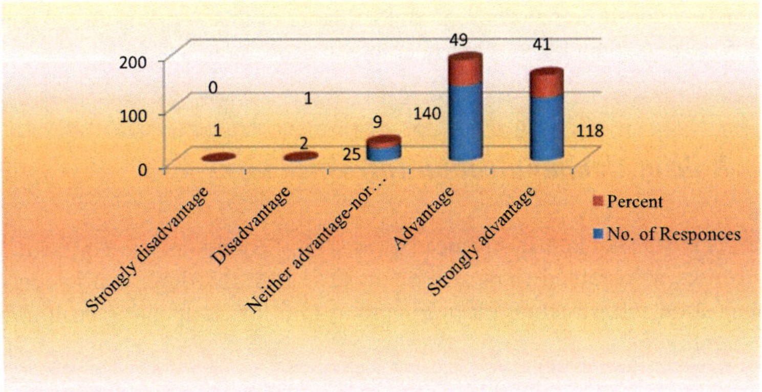

Fig. 6.5 Importance of advance production and process technology

6.2.5 Role of Advanced Production and Process Technology

With regard to next casual variable, i.e., advance factors, the role of advanced production and process technology cannot be ignored at any stage of processing. With regard to this indicator, 90% firms (49 Advantage, 41 Strongly advantage) feel it is necessary to have up to date production and processing technology, as shown in Fig. 6.5. The overall PPS score for this item was found high as 82.95%.

6.2.6 Role of IPR and Patents

MSMEs were found weak in proxy variable, i.e., IPR and patents as reflected by Fig. 6.6, which shows that only 41% firms consider it as a advantageous determinant

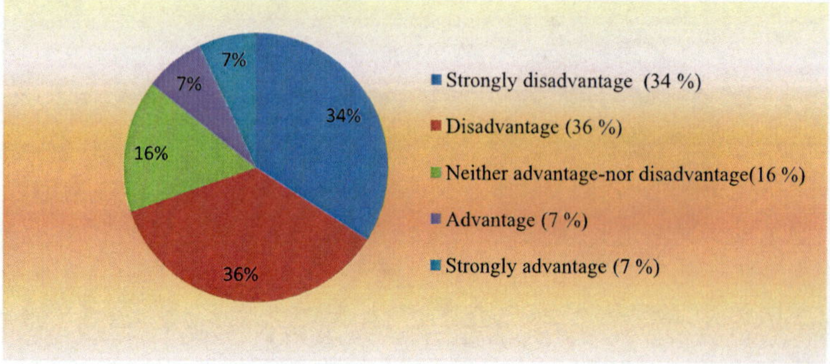

Fig. 6.6 IPR and patents

for competitive positioning. The overall PPS score of this item was found quite low with 43.57% only, which needs considerable improvement.

6.2.7 *Role of Communication Infrastructure*

With respect to the last item in advanced factors, i.e., “communication infrastructure,” survey results so obtained show that about 68% MSMEs think that having better communication infrastructure plays important role for competitive positioning, as depicted in Fig. 6.7.

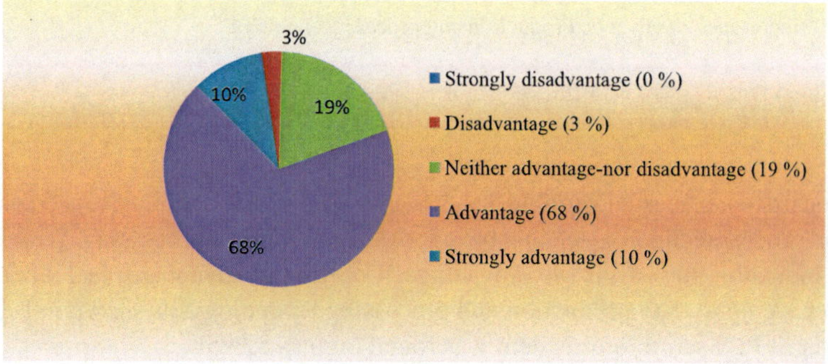

Fig. 6.7 Communication infrastructure

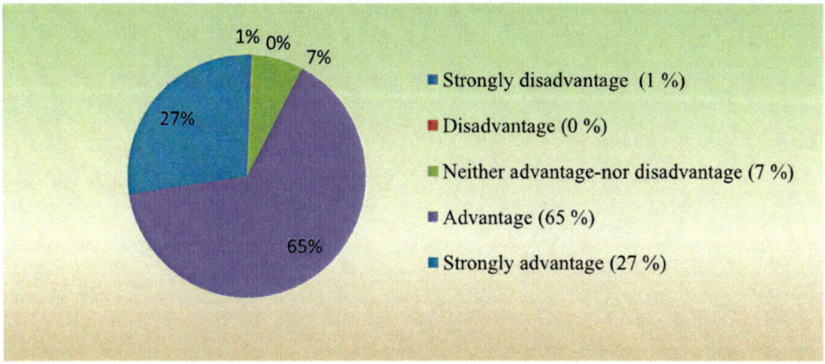

Fig. 6.8 Role of market size

6.3 Demand Conditions

The second determinant of Porter's diamond model is Demand conditions. This determinant is subdivided into two proxy variables, (i) market value and (ii) sophistication. With regard to this variable, questions were framed that gives idea about the market size, pattern of growth, distribution channels and effect of new investment in the region.

6.3.1 Role of Market Size

With respect to first item "market size," result shows that 65% firms feel it advantageous, whereas 27% feel it strongly advantageous as increase in market size helps to create competitive positioning. The overall PPS score for this item was found 83.65%. Figure 6.8 presents the details.

6.3.2 Role of Pattern of Growth

With regard to second item "pattern of growth," Fig. 6.9 shows the necessary details. It can be seen that 71% MSMEs confirm it advantageous, whereas 17% firms strongly advantageous. The overall PPS score for this item was found 81.54%, which is quite considerable. Market growth is as important as the absolute size of the market and indicates a future trend. A fast-growing domestic market inspires the firms in a country to adopt new technologies.

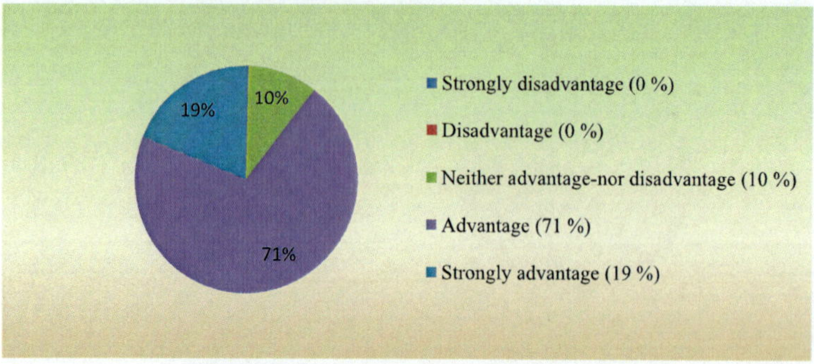

Fig. 6.9 Role of pattern of growth

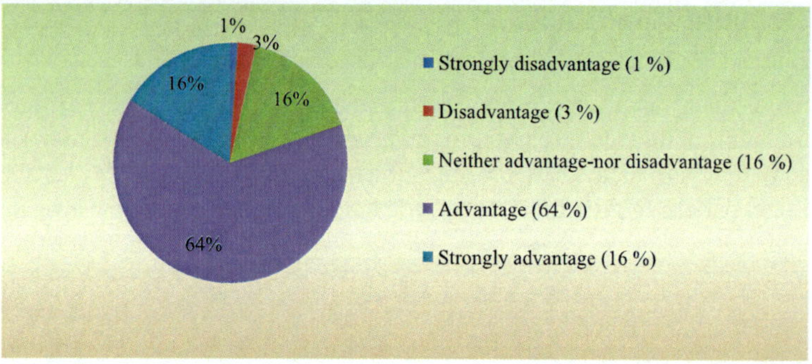

Fig. 6.10 Role of distribution channels

6.3.3 Role of Sophistication

The second casual variable, i.e., sophistication is measured using two proxy variables, i.e., distribution channels and new investment in the region. With respect to first item "distribution channel," the results are presented in Fig. 6.10. 64% firms feel it advantageous, whereas 16% find it strongly advantageous and overall PPS score is 78.37%, which shows that the distributions channel plays very important role as they directly affect the ease of availability of products to customers.

6.3.4 Role of Effect of New Investment in the Region

With regard to next proxy variable, i.e., "effect of new investment in the region," the responses so collected (Fig. 6.11) show that 65% MSMEs feel it advantageous, whereas 17% MSMEs found it strongly advantageous as it helps in creating competitive positioning. The overall PPS score for this item was found to be 79.72.

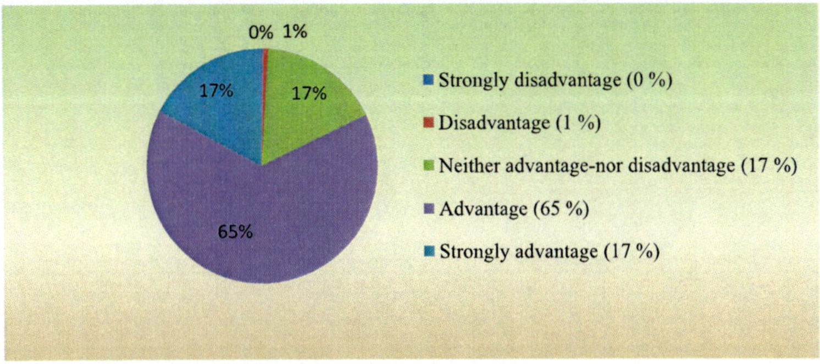

Fig. 6.11 New investment in the region

6.4 Related and Supported Industries

The third determinant of Porter's diamond model is related and supported industries. This determinant is subdivided into two casual variables, i.e., related companies and support. The first variable consists of role of technology upgradation, information flow, and share technology development and the second casual variable support consists of R&D investments, suppliers, and distribution channels and marketing.

6.4.1 Role of Technology Upgradation

With regard to the item "the role of technology up-gradation," results are presented in Fig. 6.12. It can be seen that 59% respondents agreed that it is strongly advantageous with 33% found it advantageous. The overall PPS score for this item was found high 90.00%, which shows that technology upgradation plays important role in competitive positioning of firms.

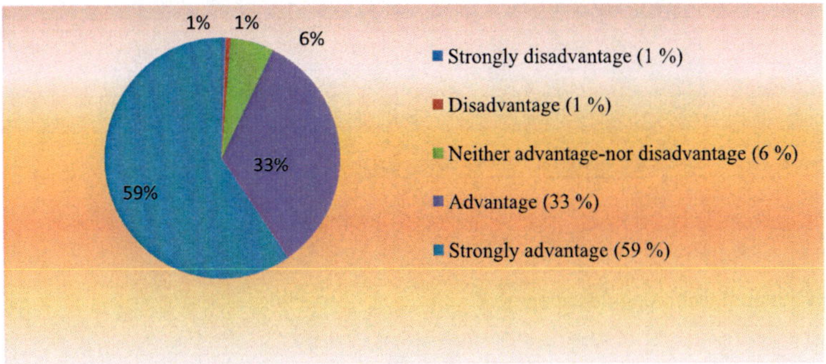

Fig. 6.12 Technology upgradations

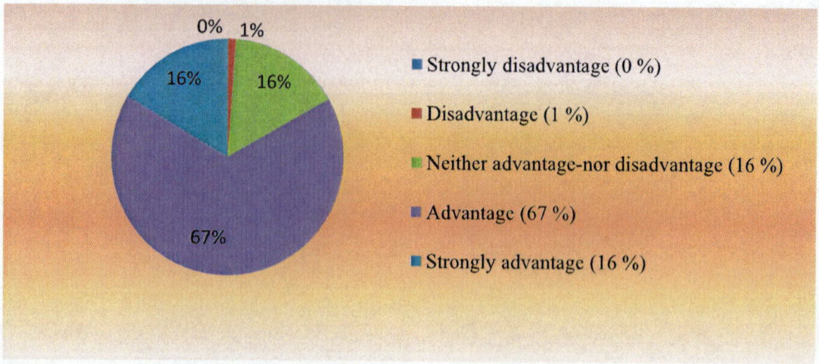

Fig. 6.13 Information flow

6.4.2 Role of Information Flow

With regard to next proxy variable, i.e., "role of information flow" results are presented in Fig. 6.13. It can be seen that 67% firms found it advantageous, whereas 16% strongly advantageous. The overall score for this item was found 79.58%.

6.4.3 Role of Financial Management

As the MSMEs are struggling with financial management and it is very important to share technology with other firms. With respect to this indicator, the responses are shown in Fig. 6.14. 55% firms describe it advantageous, whereas 32% firms describe it as strongly advantageous. The overall PPS score for this item was found 83.45.

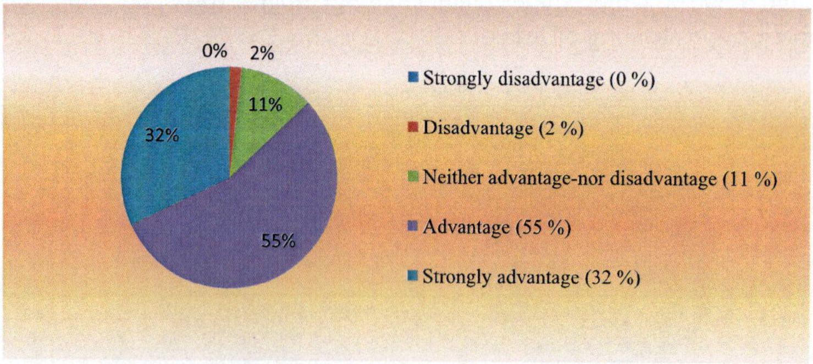

Fig. 6.14 Shared technology development

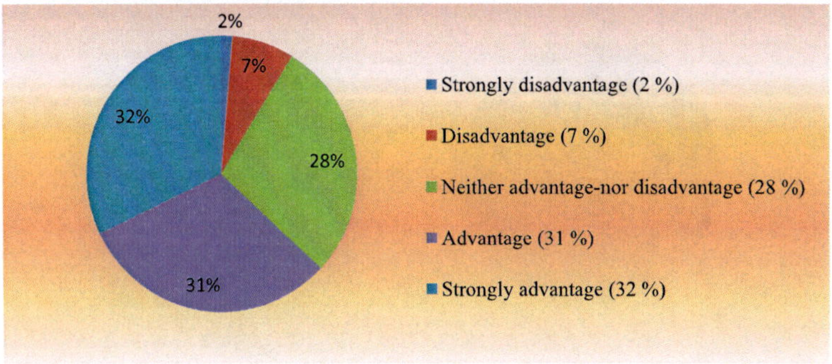

Fig. 6.15 R&D importance

6.4.4 Role of R&D Investments

With regard to "role of R&D investments," it was found that 63% firms feel it advantageous and 32% feels it strongly advantageous. The overall PPS score for this item is found 76.97%. Figure 6.15 presents the results.

6.4.5 Role of Supplier and Distribution Channel

With regard to next item "role of supplier and distribution channel," it was found that 64% firms described it advantageous whereas 15% consider it strongly advantageous. The overall PPS score for this item was found 78.46%, details are presented in Fig. 6.16.

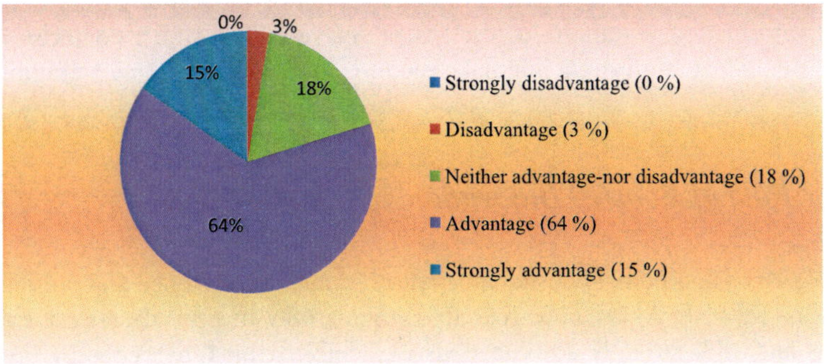

Fig. 6.16 Role of suplier and distributions channels

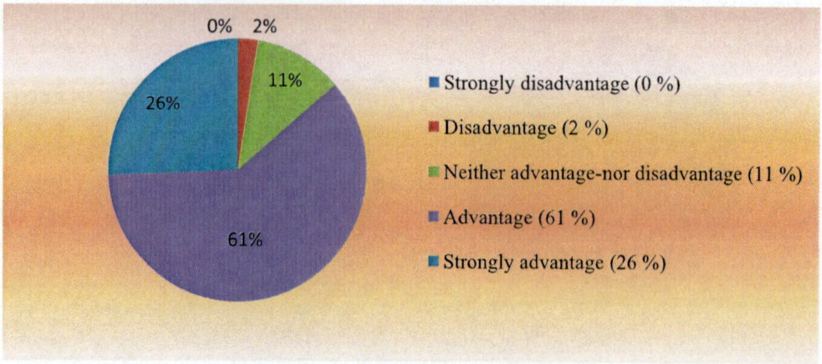

Fig. 6.17 Importance of marketing

6.4.6 Role of Importance of Marketing

The next variable is "importance of marketing," it can be seen from Fig. 6.17 that 61% firms described it advantageous where as 11% found it neither advantageous nor disadvantageous, and 26% firms market it strongly advantages. The overall PPS score for this item was found 81.89%, which shows that marketing plays important role in competitive positioning of MSMEs.

6.5 Firm Structure, Strategy, and Rivalry

The fourth Porter's determinant is firm structure, strategy and rivalry, which is further subdivided into two casual variables, i.e., rivalry and structure/strategy. Firm strategies and structure play very important role, it depends upon how firms were created, organized, and managed, the nature of domestic rivalry. Questions were framed to collect information pertaining to various proxy variables, which comprise of the rivalry, i.e., boost innovation, internal structure, and geographic concentration.

6.5.1 Role of Boosted Innovation

With regard to first item, "boosted innovation" data were obtained and presented as shown in Fig. 6.18. As it can be seen from figure, 68% of firms described it advantageous and 18% strongly advantageous to achieve competitive edge by boosting innovation. The overall PPS score for this item was found 80.56.

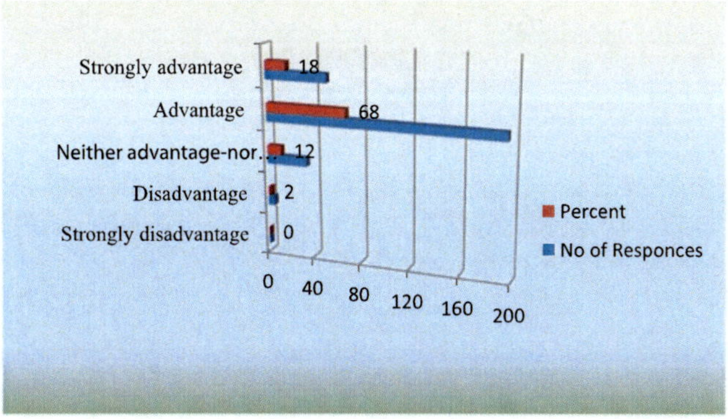

Fig. 6.18 Boost inovation

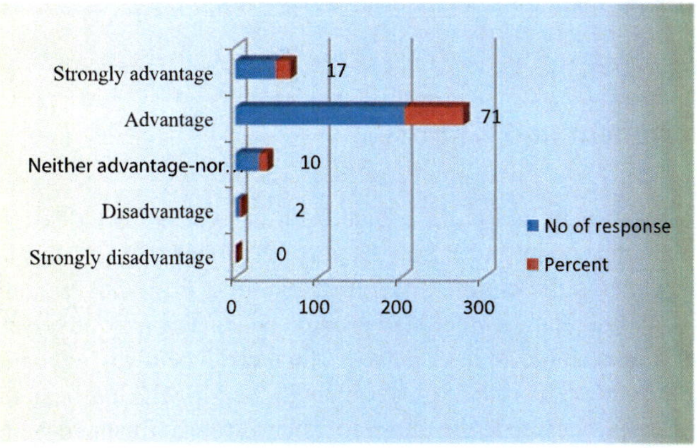

Fig. 6.19 Internal structure

6.5.2 Role of Internal Structure

With regard to the next item "internal structure," the results are presented in Fig. 6.19, it can be seen from figure that 71% firms described it advantageous whereas 17% found it strongly advantageous. The overall PPS score for this item was found 80.77.

6.5.3 Role of Geographic Concentration

With respect to next item "geographic concentration," the results are described as 49% firms found it advantageous to have geographic concentration whereas 37%

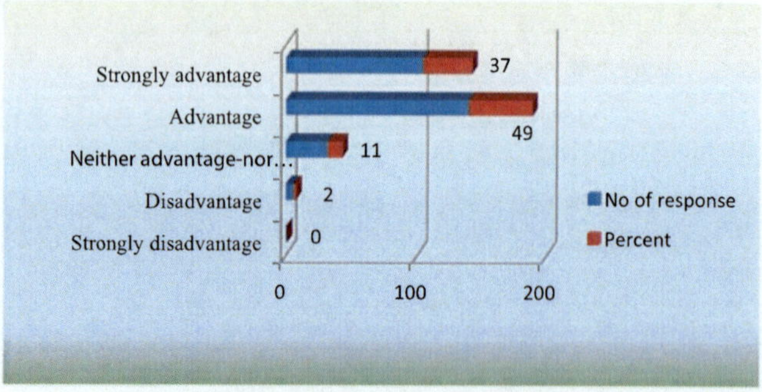

Fig. 6.20 Role of geographic concentration

with strongly advantageous as shown in Fig. 6.20. The overall PPS score was found 84.08%, which is quite high.

6.6 Government and Culture

Though the government and culture have been introduced as an external variable to Porter's diamond model but it plays a very important role and has a significant effect, which governs the competitive advantage of a firm. Government role is to make policies and regulations that may provide benefits or may adversely influence the industries. For example, subsidies, taxes, financial incentives, education policies, quality standards, etc. The role of government in the MSMEs has been rather direct. This determinant is measured with questions such as role of financial support system, environmental regulations, impact of national culture, and business climate.

6.6.1 Role of Financial Support System

With regard to first item "financial support system," shown in Fig. 6.21, 47% MSME respondents feel it strongly advantageous, whereas 42% feel it important. The overall PPS score for this determinant is found high 86.64%.

6.6.2 Role of Environmental Regulations

With regard to next item "environmental regulations," it can be seen that 67% firms find it advantageous to have proper rule and regulations regarding environment management whereas only 14% find it strongly advantageous. The overall PPS score

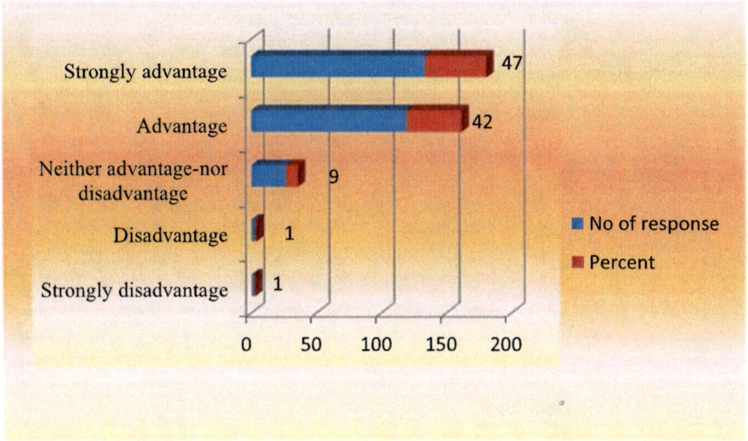

Fig. 6.21 Financial support

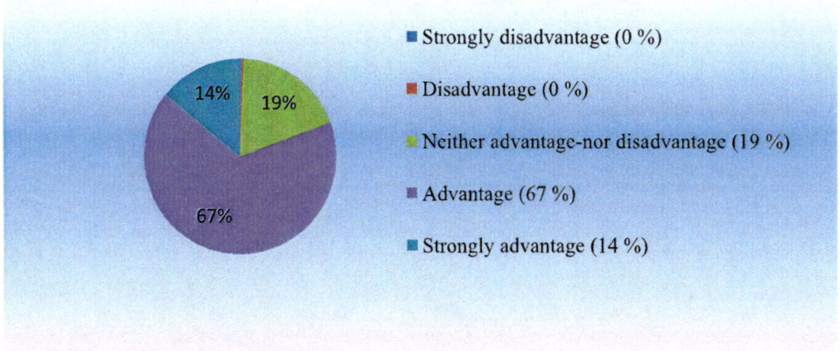

Fig. 6.22 Role of environment regulations

for this item is found 78.73. The results related to role of environment regulation on the firm competitiveness are given in Fig. 6.22 respectively.

6.6.3 Role of Impact of National Culture

The behavior of people toward quality plays an important role. The more specific demands emphasize the companies to make product with better quality, which leads to higher competitiveness. The result with respect to the proxy variable "impact of national culture" presented in Fig. 6.23 shows that 74% respondent feels it advantageous and 10% feel it strongly advantageous. The overall PPS score for this item was found 78.67.

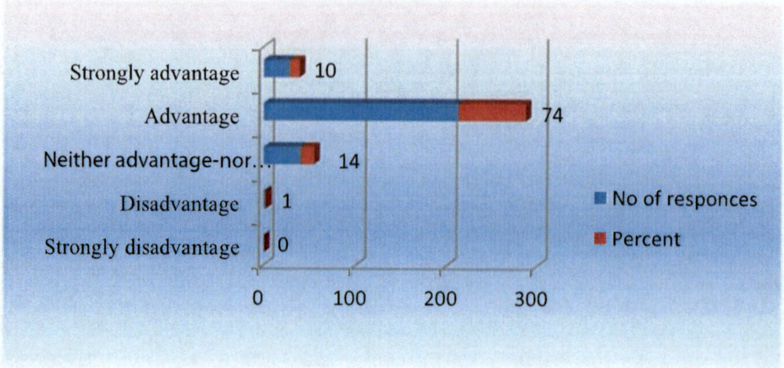

Fig. 6.23 Impact of national culture

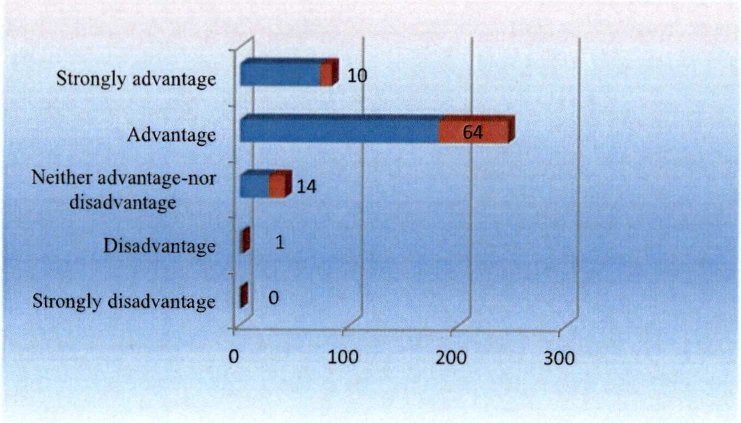

Fig. 6.24 Role of business climate

6.6.4 *Role of Business Climate*

With regard to the last item "role of business climate," shown in Fig. 6.24, it can be seen that 64% firms feel it advantageous, whereas 26% with strongly advantageous. The overall PPS score for this item was found as 82.74%.

6.7 Concluding Remarks

The analysis of each item under factor conditions reveals that IPR patents, unskilled labor, and low natural resources are some areas with low to average PPS score, need improvement. Under demand conditions, market size and pattern of growth are two prime contributors (with high PPS score) toward creating competitive advantage. With reference to status of related and supporting industries, the results found that transparency and quick information flow, and better support from suppliers and distribution channels are two main contributors with average PPS score. Whereas technology upgradations, shared technology development, relation with research development institutes, and fun allocation for R&D are critical factors, which need immediate attention to increasing technological support and consequently to creating competitive advantage. With respect to status of firm's strategy, structure and rivalry, information flow within firm has high followed by level of marketing innovation. Other items such as culture to boost innovation and knowledge regarding quality awards systems are two critical items, which need immediate attention. Under the Status of Government role, it can be seen that unfavorable labor laws and lack of allocation of funds by the government for R&D need improvement.

Chapter 7
Information on Benefits After Implementing QM Practices

Highlights

Successful implementation of quality management (QM) practices results in numerous benefits such as reduction in warranty expenditures and customers' complaints, etc. It is important to investigate the association of these benefits with firm size. Also, the casual relations among these factors are investigated using structural equation modeling approach.

7.1 Introduction

This chapter presents the details of responses (part E of sample instrument) regarding successful implementation of quality management (QM) practices in MSMEs spread across four industrial clusters of Himachal Pradesh. The survey questionnaire was designed as discussed in Chap. 3, Section E. Respondents were asked to tick on a Five-point scale (Likert) representing "2 = *Substantial increase,* 1 = *Increase,* 0 = *No change,* −1 = *Somewhat reduce and* −2 = *Substantially reduce.*" For this section, we only consider the data obtained from 215 ISO certified firms. The data were analyzed in two stages: in the first stage, descriptive analysis along with cross-tabulation and Chi-square (χ^2) test was performed to verify the association between variables and firm size. In the second stage, factor analysis was conducted by entering all variables used in analysis into the principal component analysis (varimax rotation) followed by reliability check based on Cronbach's alpha (α) value. Further, the relationships among latent variables identified after factor loading were developed using structural equational modeling (SEM).

R .K. Sharma, *Quality Management Practices in MSME Sectors*,
Springer Tracts in Mechanical Engineering,
https://doi.org/10.1007/978-981-15-9512-7_7

7.2 Benefits of QM Implementation

Descriptive Analysis

In the descriptive analysis, the status of all 16 potential indicators has been assessed. The percent point score (PPS) of each set of questions has been calculated and presented in Table 7.1. The measure of PPS score reflects the influence of different key variables on MSMEs. Further, Chi-square test with cross-tabulation was performed to test whether these potential benefits from quality management have any association with firm size. The significance of Chi-square coefficient has been evaluated based upon "*p*" value, as given in Table 7.2. Figure 7.1a–p presents the detailed results of implementing QM practices with respect to first six variables.

7.2.1 Employee Training (ET)

The actual use of quality tools, data analysis, and statistical techniques need proper training and employee's encouragement. The study results are shown in Fig. 7.1a, it has been observed that in overall almost all firms (98%) found arise in employee training. The overall percent point score for this variable was found second highest as PPS = 73.72 as evident from Table 7.1. From Table 7.2, it can be seen that the small value of Pearson's Chi-square coefficient (e.g., $\chi^2 = 2.93$, 4df), is not statistically significant as $p = 0.56$, so we can say that the effect of QM practices on employee training is not different with firm size.

7.2.2 Employee Participation (EP)

The involvements of employees in quality activities show the significance of success or failure of QM in a firm. Fig. 7.1b presents the study results showing the effects of QM practice on employee's participation. Study results found that in overall, almost all firms (96%) experienced growth in employee participation. The issue gains maximum rating with the highest value of percent point score as PPS = 75.50 as evident from Table 7.1. From Table 7.2, the slight difference within firm size was found as ($\chi^2 = 11.43$, 4df), which is statistically significant as $p = 0.02$, so we cannot reject the possibility that effects on employees participation after implementing QM practices have an association with the size of a firm.

Table 7.1 Descriptive statistics for benefits after implementing QM practices

S. no	Variables	Number of responses	Number of unit scoring					Total Point Score (TPS)[a]	Percent Point Score (PPS) $\frac{TPS}{2*N} \times 100$	Central Tendency (TPS/N)
			−2 (W_1)	−1 (W_2)	0 (W_3)	1 (W_4)	2 (W_5)			
1	Employee training and education	215	0	0	4	105	106	317	73.72	1.47
2	Employees participation	200	0	0	8	82	110	302	75.50	1.51
3	Competitive spirit	215	0	0	19	124	72	268	62.33	1.25
4	Employee morale	212	0	0	11	111	90	291	68.63	1.37
5	Quality awards	207	0	0	27	126	54	234	56.52	1.13
6	Recruitment and retaining of skilled manpower	211	0	1	14	111	85	280	66.35	1.33
7	Warranty expenditure	208	3	94	102	9	0	−91	−21.88	−0.44
8	Customer complaints	206	33	156	12	5	0	−217	−52.67	−1.05
9	Manufacturing lead time	210	8	124	57	13	8	−111	−26.43	−0.53
10	Energy consumption	207	4	50	101	49	3	−3	−0.72	−0.01
11	Maintenance cost	214	4	84	66	58	2	−30	−7.01	−0.14
12	Rework and scrap	209	30	161	10	7	1	−212	−50.72	−1.01

(continued)

Table 7.1 (continued)

S. no	Variables	Number of responses	Number of unit scoring					Total Point Score (TPS)[a]	Percent Point Score (PPS) $\frac{TPS}{2*N} \times 100$	Central Tendency (TPS/N)
			−2 (W_1)	−1 (W_2)	0 (W_3)	1 (W_4)	2 (W_5)			
13	Sales volume	213	1	3	25	155	29	208	48.83	0.98
14	Company profit	198	0	5	11	153	29	206	52.02	1.04
15	Customer relationships	201	0	3	114	78	6	87	21.64	0.43
16	Suppliers relationships	207	0	0	108	94	5	104	25.12	0.50

[a]Total Point Score (TPS) = $\mathbf{1}\times W_1+2\times W_2+\mathbf{3}\times W_3+4\times W_4+5\times W_5$

Table 7.2 Cross-tabulation results for benefits after implementing QM practices

S. no	Variables	Mean	Std. deviation	Percent Point Score (PPS)	Interaction with firm size			Chi-square test			Results
								X^2	Df	*p*-value	
1	Employee training	1.47	0.54	73.72	ET	↔	Firm size	2.93	4	0.56	✖
2	Employees participation	1.51	0.57	75.50	EP	↔	Firm size	11.43	4	0.02	✓
3	Competitive spirit	1.24	0.60	62.33	CS	↔	Firm size	5.84	4	0.21	✖
4	Employee moral	1.37	0.58	68.63	EM	↔	Firm size	1.55	4	0.82	✖
5	Quality awards	1.13	0.61	56.52	QA	↔	Firm size	3.25	4	0.53	✖
6	Recruitment and retaining	1.33	0.62	66.35	RR	↔	Firm size	25.86	4	0.00	✓
7	Warranty expenditure	−0.04	0.61	−21.88	WE	↔	Firm size	10.24	6	0.11	✖
8	Customer complaints	−1.05	0.56	−52.67	CC	↔	Firm size	3.99	6	0.68	✖
9	Manufacturing lead time	−0.53	0.82	−26.43	MLT	↔	Firm size	11.34	8	0.18	✖
10	Energy consumption	−0.01	0.78	−0.72	EC	↔	Firm size	9.41	8	0.31	✖
11	Maintenance cost	−0.14	0.87	−7.01	MC	↔	Firm size	5.34	8	0.72	✖
12	Rework and scrap	−0.01	0.61	−50.72	RS	↔	Firm size	8.35	8	0.40	✖

(continued)

Table 7.2 (continued)

S. no	Variables	Mean	Std. deviation	Percent Point Score (PPS)	Interaction with firm size			Chi-square test			Results
								X^2	Df	*p*-value	
13	Sales volume	0.97	0.59	48.83	SV	↔	Firm size	6.45	8	0.65	✖
14	Company profit	1.13	0.55	52.02	CP	↔	Firm size	3.45	6	0.75	✖
15	Customer relationships	0.43	0.58	21.64	CR	↔	Firm size	5.63	6	0.47	✖
16	Suppliers relationships	0.50	0.55	25.12	SR	↔	Firm size	22.66	4	0.00	✓

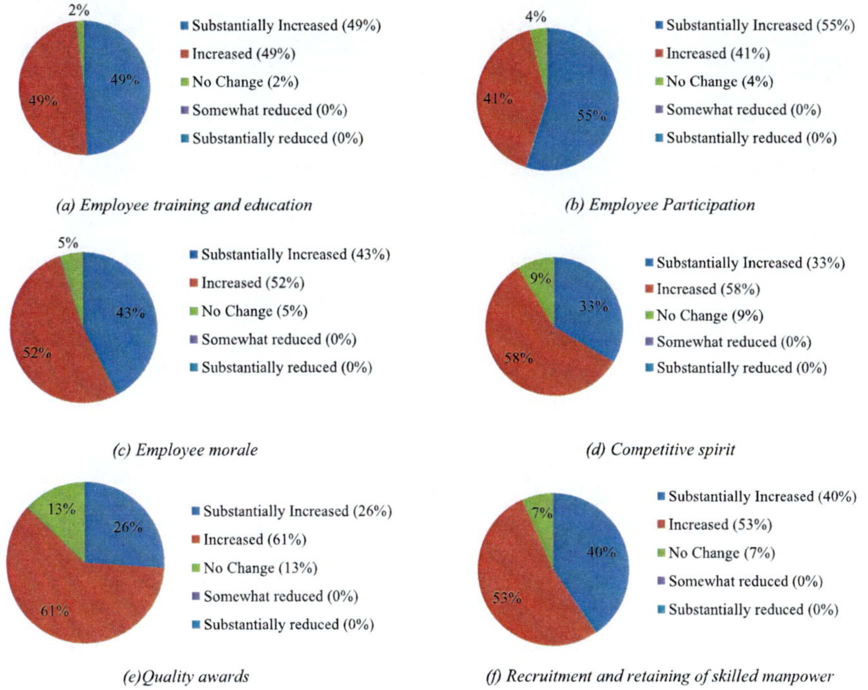

Fig. 7.1 **a–f** Effect of QM practices on identified variables. **g–l** Effects of QM practices on identified variables. **m–p** Effect of QM practices on identified variables

7.2.3 *Effect of Employee Morale (EM)*

High employee morale is a sign of quality at a workplace. Motivation can be seen from the certain psychological characteristic, which enables the individual to behave in a particular form to accomplish a specific goal. Study results presented in Fig. 7.1c show that overall 43% firms noted substantial increase, and 52% firms feel a slight increase in employee morale. The overall percent point score was found quite sound as PPS = 68.63 as evident from Table 7.1. From Table 7.2, the slight difference of results within firm size is not found statistically significant as shown by the small value of Chi-square co-efficient ($\chi^2 = 1.55$, 4df), which is not statistically significant as $p = 0.82$, so we can say that employee morale increased irrespective of firm size.

7.2.4 *Effect on Competitive Spirit (CS)*

Quality management develops committed and competent employees, who are loyal to enhance firm performance. Competitive spirit raises the motivation toward quality

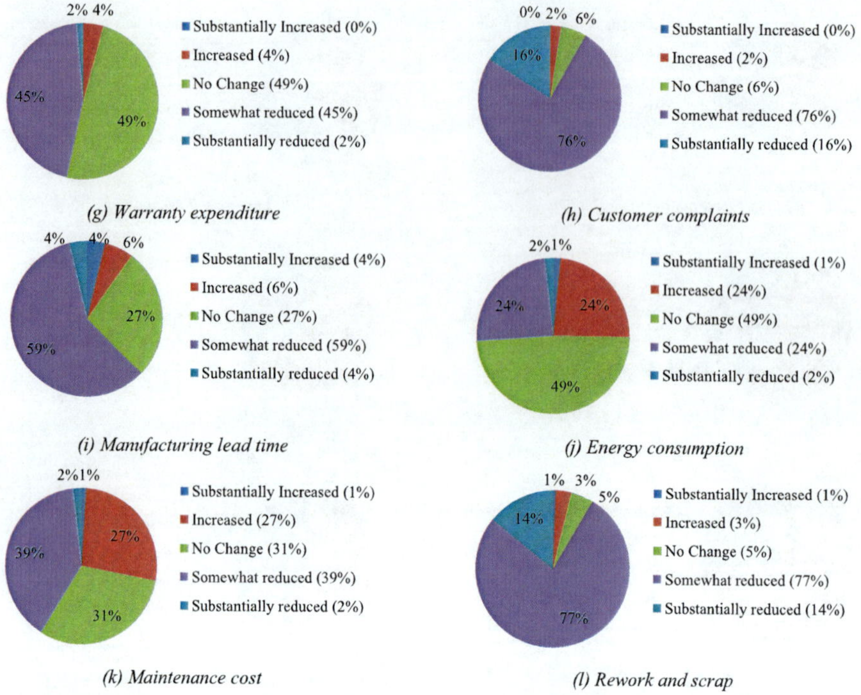

Fig. 7.1 (continued)

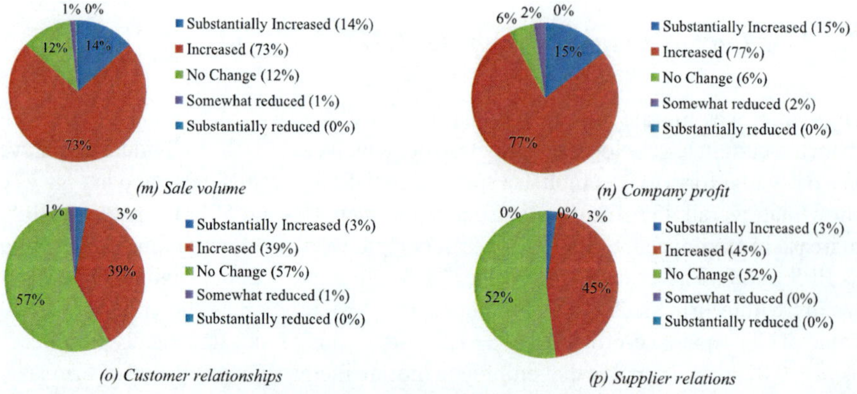

Fig. 7.1 (continued)

implementation and vice-versa, which leads to high productivity and profitability of the firm. More than 90% firms observed a surge in the competitive spirit of employees after implementing QM practices (Fig. 7.1d). The present point score was found quite well as PPS = 62.33 as evident from Table 7.1. It shows the overall importance of competitive spirit. From Table 7.2, the value of Chi-square coefficient was found small as ($\chi^2 = 5.84$, 4df), which is not statistically significant as $p = 0.21$, so we can reject the possibility that there is any association between firm size and competitive spirit.

7.2.5 *Quality Awards (QA)*

A well-organized firm must develop a formal system for the encouragement of individual, team and any section or department of the company. After implementing QM practices, 26% firms noted substantially hike and 61% slightly increase in rewarding that might involve increments, monetary awards, and some organizations may have profit-sharing program as evident in Fig. 7.1e. The overall percent point score (PPS) for this aspect of the study was found suitable as PPS = 56.52 as evident from Table 7.1. From Table 7.2, the slight different within firm size is not found statistically significant as ($\chi^2 = 3.25$, 4df), at $p = 0.53$. So, we can say that companies adopting QM practices have a systematic awarding and recognition system irrespective of firm size.

7.2.6 *Recruitment and Retaining (R&R)*

After implementing QM practices, firms have been able to strengthen their HRM process (PPS = 66.35). Almost all firms (93%) (Fig. 7.1f) experience changes in recruitment and hiring process. The overall percent point score was considered good as PPS = 66.35 as evident from Table 7.1. There is a sufficient difference in results concerning firm size. The results found a high value of Chi-square coefficient e.g., ($\chi^2 = 25.86$, 6df), (Table 7.2) which is statistically significant at $p < 0.001$, so we cannot reject the possibility that there is any association between firm size and effect on recruitment and retaining of employees.

7.2.7 *Warranty Expenditure (WE)*

For the establishment of a long-term relationship with customers and to create market value, firms need to provide product's quality certificate and attractive warranty policies. With this aspect of the study, we found a total of 208 valid responses, which consist of 41 (20%) micro, 100 (48%) small and 67 (32%) medium firms. From the results of the study, it has been observed that overall 49% firms found no change in warranty cost whereas 45% feel a somewhat reduction in the cost which they spend on the warranty of their product(s) as shown in Fig. 7.1g. The negative score in overall percent point score, e.g. PPS = − 21.88 (as evident from Table 7.1) indicates that warranty expenditures can be reduced by implementing QM practices. The slight different with respect to firm size was noted and which is not statistically significant as depicted (From Table 7.2) from a low value of Chi-square co-efficient ($\chi^2 = 10.24$, 6df, $p = 0.115$).

7.2.8 *Customer Complaints (CC)*

Direct and open communications with customers allow companies to identify their requirements and further it helps to estimate what improvements can be made? From Fig. 7.1h, it can be seen that overall 76% firms observed slight reduction and 16% firms feel a substantial reduction in customer complaints after implementing QM practices. The percent point score for this variable was found highest in negative as PPS = − 52.67 as evident from Table 7.1. From cross-tabulation of data (Table 7.2), it was concluded that there is no significant statistical association between firm size and effect of QM practices on customer complaints as indicated by the low value of Chi-square coefficient as ($\chi^2 = 3.99$, 6df), which is not statistically significant as $p = 0.68$.

7.2.9 *Manufacturing Lead Time (MLT)*

QM practices apply a structured approach to manage activities, which improve process design activities. From detailed results in Fig. 7.1i, it can be seen that more than half of firms (63%) observed a reduction in manufacturing lead time. The overall PPS score was found PPS 26.43 (as evident from Table 7.1) shows a decrease in total manufacturing lead-time. Study results (Table 7.2) noted that there is no statistically significant difference with respect to firm size as indicated by the low value of Chi-square coefficient as ($\chi^2 = 11.34$, 8df), which is not statistically significant as $p = 0.183$, so we can say that QM tools and techniques affect equally irrespective of firm size.

7.2.10 *Energy Consumption (EC)*

The study found mixed results for this aspect as 49% firm is observed no changes, whereas around 24% firms observed somewhat increase and the same percentage said somewhat reducing energy consumption as shown in Fig. 7.1j. The percent point score of this variable was found as PPS = −0.72 (as evident from Table 7.1) The results shown in Table 7.2 of cross-tabulation are found as ($\chi^2 = 9.41$, 8df), which is not statistically significant as $p = 0.31$, so we can say that QM practices affect equally with respect to energy consumption irrespective of firm size. During personal interviews, it was noted that, at starting phase of QM program, firms noted an increase in energy consumptions, but it started reducing after some time.

7.2.11 *Maintenance Cost (MC)*

With an intention to fulfill the requirements of sophisticated automatic systems, the core types of maintenance have been extended from preventive, predictive and corrective to detective maintenance. With regard to this aspect of the study, it was found that overall 41% firms feel a reduction in maintenance cost and 31% firms found no change, whereas 27% firms recorded increase in total maintenance quality cost as shown in Fig. 7.1k. The overall percent point score was found in negative as PPS = −7.01 as evident from Table 7.1. The effect of QM practices was obtained even for all size of firms as indicated by the small value of Chi-square coefficient as ($\chi^2 = 5.34$, 8df), which is not statistically significant as $p = 0.721$ as shown in Table 7.2.

7.2.12 *Rework and Scrap (RS)*

Quality management promotes preventive maintenance to increase machine reliability, which consequently reduces the rework and scrap. Study results as shown in Fig. 7.1l found almost all firms (91%) experienced a reduction in rework and scrap. The overall percent point score (PPS) was considered highly negative as PPS= −50.72 as evident from Table 7.1. The negative value of PPS score shows the positive impact of successful implementation of statistical tools and techniques. Further, from Table 7.2 results, it is noted that there is no significant association between the firm size and reduction in rework and scrap as indicated by the small value of Chi-square coefficient ($\chi^2 = 8.35$, 8df). It is not statistically significant as $p = 0.41$; it depicts that QM practices are equally important for all size of MSMEs firms. During personal interviews, it has been observed that the initial quality level of these companies was so low that a fraction of success in implementing QM practices would drastically improve their quality level.

7.2.13 *Sales Volume (SV)*

With this regard, overall 73% firms feel increase in sales volume, whereas 14% firms found with a substantial increase in sales volume as shown in Fig. 7.1m. The percent point score for this aspect of the study was determined as PPS= 48.83 as evident from Table 7.1. From the low value of Chi-square coefficient (From Table 7.2), it can be stated that QM practices' effect equally to all size of firms, as indicated by (χ^2= 6.45, 8df), which is not statistically significant as $p = 0.65$.

7.2.14 Company Profit (CP)

A total of 198 MSMEs firms replied to this aspect of the study. Overall 77% firms admitted that there is an increase in company profit after implementing QM practices as shown in Fig. 7.1n. Percent point score for this aspect of the study was found suitable as PPS = 52.02 as evident from Table 7.1. Results from cross-tabulation indicate that QM affects equally in this aspect irrespective of firm size as shown by the low value of Chi-square coefficient ($\chi^2 = 3.45$, 6df), (Table 7.2) which is not statistically significant as $p = 0.75$.

7.2.15 Customers Relations (CR)

In the course of implementing QM, firms start involving customers in various activities PPS = 21.64 as evident from Table 7.1. Around half of the companies (42%) start inviting customers to visit their plant, which helped to collect the detailed information about their needs and expectations as shown in Fig. 7.1o. The managers, which have direct contact with clients can readily, obtain first-hand information about products, service quality, and use such information in the making firm's strategic decisions. The small value of Chi-square co-efficient ($\chi^2 = 5.63$, 6df), which is not statistically significant as $p = 0.47$, shows that there is no association between firm size and customer relations e.g. it affects equally to all irrespective of firm size (Table 7.2).

7.2.16 Supplier Relationship (SR)

During implementing QM, firms have started maintaining their database regarding raw material quality, delivery time and pricing, etc. Study results found that overall 48% firms feel a rise in relations with suppliers during QM implementation as shown in Fig. 7.1p. The overall percent point score for an aspect of the study was obtained as PPS=25.12 as evident from Table 7.1. Further, the value of Chi-square coefficient (Table 7.2) was found high as ($X^2 = 24.661$, 4df), which is statistically significant at $p < 0.001$, it shows that suppliers relation has a significant association with firm size.

7.3 Structural Equation Model for Relationship Among Variables

Structural equation modeling is a multivariate statistical analysis technique that is used to analyze structural relationships. It is also called causal modeling because it tests the proposed causal relationships. This technique is the combination of factor analysis and multiple regression analysis, and it is used to analyze the structural relationship between measured variables and latent constructs. To determine the relationship among the benefit variables, exploratory factor analysis was implemented using principal component and Varimax with Kaiser Normalization as rotation method on 16 QM indicators. Only those items that accounted for variance >1 (Eigen-value >1) were extracted. The results of KMO test were found adequate as 0.79, which is higher than 0.60, which indicates sufficient intercorrelation and Bartlett's test of sphericity was considered significant as ($\chi^2 = 1055$, 120df, $p < 0.000$). The next step was to perform factor loading, as per Conca et al. (2004); Brkic et al. (2008), loading of each item must be greater than 0.5 and should not have any significant cross-loadings. Further, the relationships among latent variables are identified after factor loading were developed using structural equation modeling (SEM).

Based on the acceptable results of factor loading and pattern matrix, we reduced all 16 variables, three principal component factors. The first principal component explained nearly 25% of the total variance. Thus, the common method bias is not a serious issue (Podsakoff et al. 2003). Table 7.3 summarizes the results of factor analysis.

Reliability and Validity Check

After establishing pattern matrix, we opted for reliability and validity check. Table 5.19 presents the results of reliability and validity checks. Cronbach's (α) for three individual latent variables ranged from 0.857 to 0.913 indicating a high reliability of the instrument. The values of average variance explained (AVE > 0.5) and composite reliability (CR > 0.7) indicated high-scale validity. Further, the values of maximum shared variance (MSV) and average squared shared variance (ASV) indicated discriminant validity (MSV < ASE > ASV) of scale used.

Classification of Benefits Based on Factor Loading

First component shows the effects of quality management practices on six aspects, e.g., (i) employee training (ET), (ii) employee participation (EP), (iii) competitive spirit (CS), (iv) employee morale (EM), (v) rewards and recognition (RR), and (vi) recruitment and retaining (RSM) of skilled workers. Second component also consists of six variables, which shows the effects on (i) warranty expenditure (WE), (ii) customer complaints (CC), (iii) manufacturing lead-time (MLT), (iv) energy consumption (EC), (v) maintenance costs (MC), and (vi) rework and scrap (R&S).

Table 7.3 Related factors loading

S. no		1	2	3	4	5	6	7	8	9	10	11	12	13	14	15	16
Items		ET	EP	CS	EM	QA	RR	WE	CC	MLT	EC	MC	R&S	SV	CP	CR	SR
Components	1	**0.822**	**0.854**	**0.817**	**0.801**	**0.821**	**0.546**	0.015	−0.142	−0.081	0.010	−0.440	−0.039	0.150	0.083	−0.102	0.022
	2	−0.055	−0.037	−0.185	−0.093	−0.050	0.080	**0.701**	**0.641**	**0.878**	**0.542**	**0.576**	**0.681**	−0.052	−0.056	−0.070	0.010
	3	−0.007	−0.080	0.013	−0.014	0.073	0.168	0.059	−0.203	−0.301	−0.112	0.093	0.092	**0.882**	**0.885**	**0.937**	0.903

Note Extraction method: principal component analysis. Rotation method: Varimax with Kaiser normalization

Similarly, the third component consists of four variables they are (i) sales volume, (ii) company profit (SV), (iii) customer relations (CF), and (iv) supplier relations (SR).

- The first group of variables particularly related to participation of employees, staff, teamwork, recruitment and selection procedure, policies regarding training and educations and rewards and reorganizations. According to literature Boselie and Wiele (2000); these variables can be grouped into one latent variable (construct) and can be studied as the overall effect on firm's human effectiveness (HRM).
- The variables under second principal component are found related to the cost because of poor quality. So, as per literature Sachdeva et al. (2008), Kumar and Sosnoski (2009), Sadikoglu and Zehir (2010), Su et al. (2015), these variables can be grouped into one latent variable and can be studied as the effect of QM practices on *Quality Cost (QC)*.
- The third group of variables (e.g., third principal component) shows the effect of QM practices particularly in term of sales volume, customer's relationships, and distribution channels. As per the literature Agus et al. (2008); Gupta et al. (2006); Prajogo and Sohal (2006); these variables can be grouped into a single latent and can be studied as the overall effect on the *firm performance.*

Based on the empirical analysis, we found the three primary major areas of benefits after implementing quality management practices, i.e., *HRM effectiveness, quality cost and market performance of the firm.*

Structural Models

The relationship between identified latent variables was established by constructing a structural model using structural equation modeling (SEM). The SEM models in Fig. 7.2 simultaneously representing the casual relation between HRM effectiveness, QC, and firm performance, where circles represent latent variables and rectangle depict constituting items. Table 7.4 presents the path coefficient (β), associated t-valve, significance level (*p*-value), and standard error (S.E).

The path-I from HRM effectiveness to firm performance showing the positive association ($\beta = 0.83$, $t = 6.961$, $p < 0.01$), it states that HRM effectiveness has a direct relationship with firm performance. These results support previous findings of Thorpe et al. (2009); Dubey et al. (2015). Similarly, the path-II from QC to firm performance ($\beta = -0.33$, $t = -2.855$, $p < 0.01$) shows that QC is negatively related to firm performances. In another word, study results found that if cost due to poor quality increases, firm's performance decreases. These findings support the previous finding by Kumar and Sosnoski (2009).

At last from path-III, it is observed that a negative relationship between HRM effectiveness and QC. The path coefficient between HRM effectiveness and QC ($\beta = -0.11$, $t = -1.966$, $p < 0.05$) shows that if a firm promotes human resource management, quality cost decreases (Table 7.5). The results of fit indices for overall measurement model were estimated. Various fit such as normed Chi-square = 2.97, root mean square error of approximation (RMSEA = 0.07), goodness of fit (PGFI =

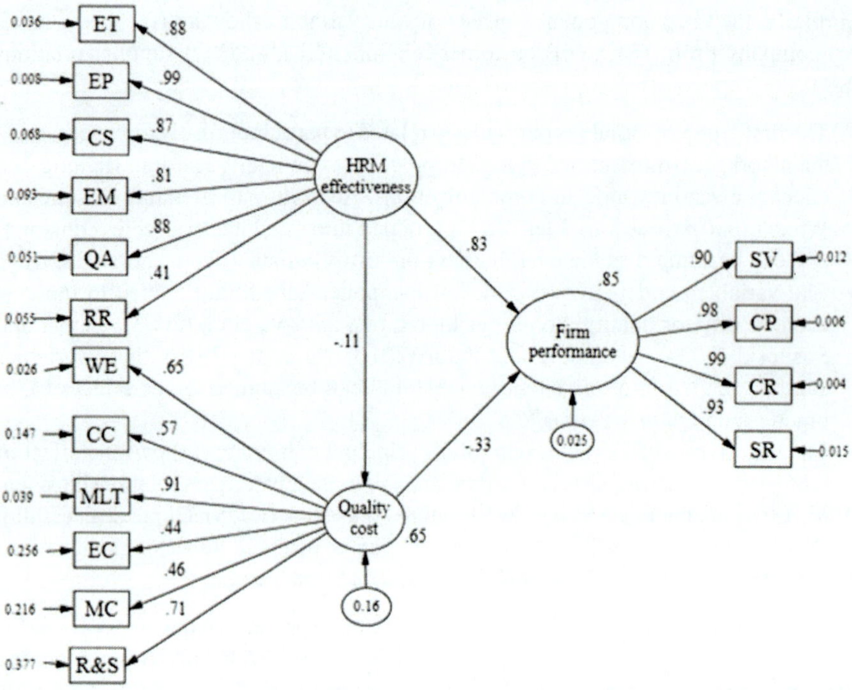

Fig. 7.2 Structural model for the relationship between HRM effectiveness, QC and firm performance

Table 7.4 Results of measurement properties for HRM effectiveness, quality cost, and firm performance

Path				Sign	Standardized regression weights (β)	*t*-value	Standard error (S.E)	R^2
	Relationship between							
Path 1	HRM effectiveness	→	Firm performance	+	0.83	6.961***	0.467	0.84
Path 2	Quality costs	→	Firm performance	−	0.33	−2.855***	0.067	
Path 3	HRM effectiveness	→	Quality costs	−	0.11	−1.966**	0.124	0.65

Note $**p < 0.05$; $***p < 0.01$

0.52), normed fit index (PNFI = 0.64), and comparative fit index (CFI = 0.94). All fit indices satisfied the recommended cut-off values (Kaynak 2003) and demonstrated that the model is reliable and strongly supported by the data collected.

Table 7.5 Results of Path model for HRM effectiveness, quality cost and firm performance

Latent constructs		Correlation coefficient (r)		Cronbach's alpha (α)		Average variance extracted (AVE)	Composite reliability (CR)	Maximum shared variance (MSV)	Average shared variance (ASV)
1	HR effectiveness	1.00			0.866	0.614	0.904	0.129	0.059
2	Quality cost	−0.11	1.00		0.857	0.503	0.857	−0.048	−0.030
3	Firm performance	0.36	−0.22	1.00	0.913	0.814	0.946	0.129	0.041

Criteria's to establish Reliability $\alpha > 0.7$; Convergent validity $0.5 < AVE < CR > 0.7$; Divergent validity $MVS < AVE > ASV$

7.4 Concluding Remarks

- Firms have started involving their employees in various planning and decision-making activities (PPS = 75.50). In almost all companies (96%) employees have begun participation in general QM activities. For achieving the higher quality goals, firms have started organization training programs at different level of specialization (PPS = 73.72). These training programs motivated the employees (PPS = 68.63) to accomplish a particular goal. Almost all units (95%) observed psychological changes in the employee's behavior. Encouraged employees started feeling ownership of their jobs and striving for personal commitments toward quality (PPS = 62.33). Furthermore, study results observed there is an increased in rewards and organization (PPS = 56.32), and recruitment and hiring process in firms (PPS = 66.35).
- After implementing QM practices, firms were able to reduce the warranty expenditures (PPS = −21.88), and customers complaints (PPS = −52.67). Most of the firms (76%) started open communication with clients to identify their real requirements. Almost 39% companies noted increased in customer relation have begun involving customers into various quality activities. Overall 77% companies observed growth in company profits (PPS = 52.02). In the course of improving customer relations, firms have realized the importance of better material flow, delivery time and pricing (PPS = 25.12). Almost half (45%) companies observed increased in supplied relation after implementing QM practices.
- With the application of seven basic QC tools like cause and effect diagrams, Pareto charts, flow charts, etc., firms observed a reduction in manufacturing lead time (PPS = 26.43). With the use of standardized components and mistake-proof equipment, companies can reduce scrap (PPS = −50.72). Most of the units (77%) experienced a somewhat reduction in rework and scrap. These improvements in process capability make a reduction in energy consumptions (PPS = 0.72).
- The study results found that only 3 variables out of 16, e.g., (i) employee's participation, (ii) recruitment and retaining of skilled employees, and (iii) effect on suppliers relationships have an association with firm size.
- Based on exploratory factor analysis, the 16 QM outcome indicators classified into three latent variables, i.e., (i) effect of human resource management, (ii) effect of the cost of poor quality, and (iii) effect on marketing performance of the firm. From this study result, it can be observed that there is no unique model, which shows all benefits from QM practices, but they can be grouped respective into latent variables such as (i) HRM effectiveness, (ii) Quality costs (QC), and (iii) firm performance.
- The results conclude that HRM effectiveness has a positive relationship with firm performance as well as a negative association with QC. Thus, an emphasis on QM is likely to provide advances in HRM effectiveness, the decline in QC and consequently improves in overall firm's performance.

References

Agus A (2011) Enhancing production performance and customer performance through total quality management (TQM): strategies for competitive advantage. Procedia-Soc Behav Sci 24:1650–1662

Agus A, Hassan Z, Noor ZM (2008) A perspective of customer relation management in supply chain management in enhancing performance. SEARC: Globalization. Innovation and Leadership, Kuala Lumpur, Malaysia

Boselie P, Wiele VDT (2002) Employee perceptions of HRM and TQM and the effect on satisfaction and innovation to leave. Managing Serv Qual 12(3):165–173

Brkic VK, Djurduja T, Dondur N, Klarin MM, Tomic B (2013) An empirical examination of the impact of quality tools application on business performance: evidence from Serbia. Total Qual Manag 24(5):607–618

Conca FJ, Llopis J, Tari JJ (2004) Development of a measure to assess quality management in certified firms. Eur J Oper Res 156(3):683–697

Dubey R, Gunasekaran A, Ali SS (2015) Exploring the relationship between leadership, operational practices, institutional pressures and environmental performance: a framework for green supply chain. Int J Prod Econ 160:120–132

Gupta S, Tewari PC, Sharma AK (2006) TPM concept and implementation approach. Maintenance World 21:1–18

Kaynak M (2003) The relationship between total management practices and their effect on firm's performance. J Oper Manag 21(6):405–435

Kumar S, Sosnoski M (2009) Using DMAIC six sigma to systematically improve shop floor production quality and costs. Int J Product Perform Manag 58(3):254–273

Podsakoff PM, MacKenzie SB, Lee JY, Podsakoff NP (2003) Common method biases in behavioral research: a critical review of the literature and recommended remedies. J Appl Psychol 88(5):879

Prajogo DI, Sohal AS (2006) The relationship between organization strategy, total quality management (TQM), and organizational performance-the mediating role of TQM. Europ J Oper Res 68(1):35–50

Sachdeva A, Kumar D, Kumar P (2008) A methodology to determine maintenance criticality using AHP. Int J Prod Qual Manag 3(4):396–412

Sadikoglu E, Zehir C (2010) Investigating the effects of innovation and employee performance on the relationship between TQM practices and firm performance: an empirical study of Turkish firms. Int J Prod Econ 127(1):13–26

Su HC, Dhanorkar S, Linderman K (2015) A competitive advantage from the implementation timing of ISO management standards. J Oper Manag 37(1):31–41

Su Q, Li Z, Zhang SX, Liu YY, Dang JX (2008) The impacts of quality management practices on business performance: an empirical investigation from China. International Journal of Quality & Reliability Management 25(8):809–823

Thorpe R, Cope J, Ram M, Mike P (2009) Leadership development in small- and medium-sized enterprises: the case for action learning' action learning. Res Pract 6(3):201–220

Chapter 8
Critical Success Factors in Adoption of QM Practices

Highlights

- Critical success factors (CSFs) that may help MSMEs in successful adoption of quality management practices are classified into three categories, i.e., (i) Strategic factors, (ii) Tactical and (iii) Operational factors.

8.1 Introduction

This chapter presents the details about the critical success factors (CSFs), which may help MSMEs in successful adoption of quality management practices. Strategic factors are subdivided into nine subfactors, tactical factors are subdivided into seven subfactors and operational factors are subdivided into six subfactors. Respondents were asked to tick on CSFs, which may help MSMEs in successful adoption of Quality Management Practices on 1–5 scale with 1 = not important, 2 = some-what important, 3 = moderately important, 4 = important, 5 = very important. Table 8.1 presents the responses obtained after administration of survey instrument in four industrial clusters of the state of H.P.

8.2 Strategic Factors

With respect to "strategic factors," there are total of nine items that represent the importance of strategic factors.

R .K. Sharma, *Quality Management Practices in MSME Sectors*,
Springer Tracts in Mechanical Engineering,
https://doi.org/10.1007/978-981-15-9512-7_8

Table 8.1 Critical factors for implementing QM practices

Sr. no	Factors	Sub criterions	1	2	3	4	5	PPS
1	Strategic factors	Top management commitment	0	1	3	87	215	93.73
		Quality culture	0	2	11	193	78	84.44
		Quality systems	1	0	11	206	68	83.78
		Quality awards	0	1	11	134	139	88.84
		Continuous improvement and innovation	0	1	12	177	95	85.61
		Benchmarking	1	4	60	153	53	78.67
		Technical expertise	1	0	7	136	142	89.23
		Marketing expertise	1	1	17	179	85	84.45
		Financial access	1	20	52	93	120	81.75
2	Tactical factors	Employee encouragement	1	0	4	139	142	89.44
		Employee involvement	1	0	5	142	136	89.01
		Training and education	1	0	5	91	188	92.63
		Teamwork	1	0	6	148	130	88.49
		Information and analysis	0	0	16	197	71	83.87
		Supplier management	0	5	42	188	51	79.93
		Communication	0	2	19	178	86	84.42
3	Operational factors	Product and service design	1	2	68	174	39	77.46
		Process management	1	2	71	166	45	77.68
		Customer focus	0	2	5	81	195	93.14
		Human resource management	0	0	7	129	150	90.00
		Industrial relations	1	1	7	205	71	84.14
		Employee rewards and incentives	0	1	8	146	131	88.46

Where 1 = Not important, 2 = Some-what important, 3 = Moderately important, 4 = Important, 5 = Very important

8.2.1 Top Management Commitment

To begin, with first item "top management commitment," detailed survey results are presented in Fig. 8.1. It can be seen that 75% respondents said it is very important aspect for successful implementation of quality management practices in firms, whereas 23% firms confirm it as important dimension. The overall PPS score for this aspect was found 93.73% respectively.

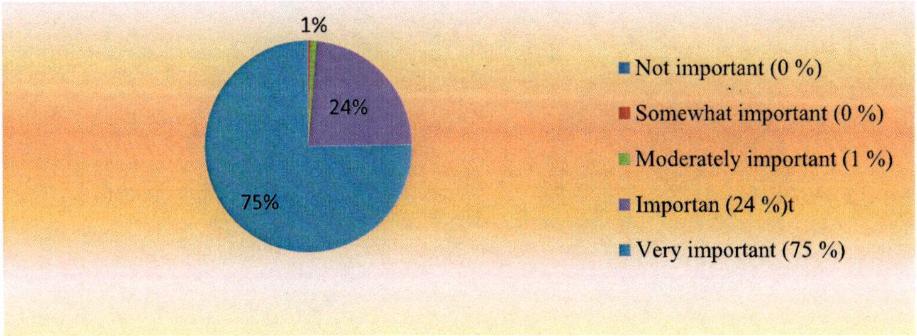

Fig. 8.1 Role of top management

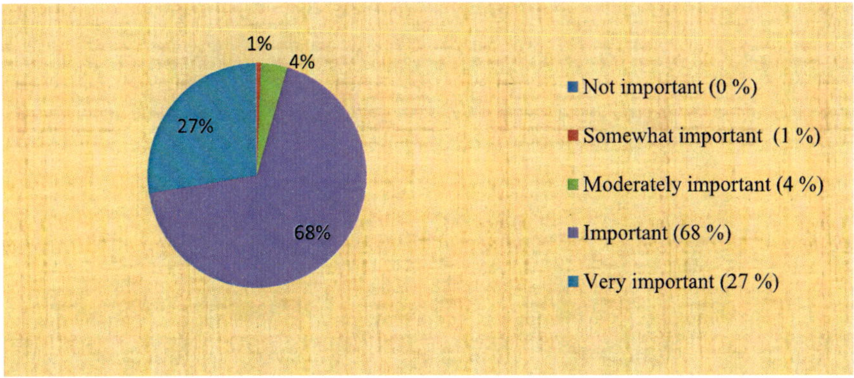

Fig. 8.2 Role of quality culture

8.2.2 Quality Culture

With respect to the second point "quality culture," responses from MSMEs respondents are presented in Fig. 8.2. It can be seen that 68% firms feel it important, whereas 27% said it is very important aspect of implementing quality practices in the firm. The overall PPS score for this point was found 84.44%.

8.2.3 Quality Systems

With regard to third point "quality systems," Fig. 8.3 represents the survey results. It was found that 72% respondents said quality system is important, whereas 24% said it is very important aspect for implementing quality practices. These two contribute toward PPS score 83.78%.

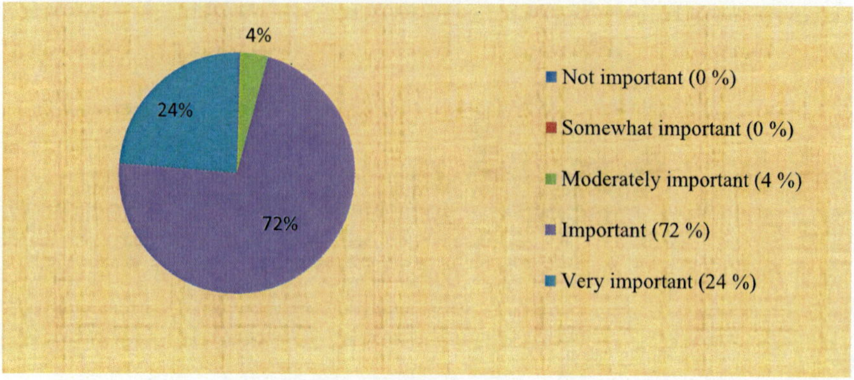

Fig. 8.3 Role of quality system

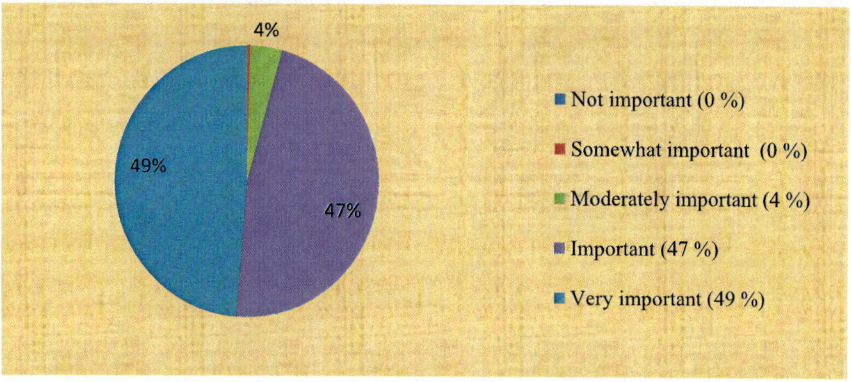

Fig. 8.4 Quality awards

8.2.4 Quality Awards

With respect to "quality awards," survey results found that 49% respondents said it is very important whereas 47% tick on important. The overall PPS score for this point was found 88.84%, which is quite high. The details are presented in Fig. 8.4 respectively.

8.2.5 Continuous Improvement and Innovation

Survey results for the factor "continuous improvement and innovation" are presented in Fig. 8.5. It can be seen that 62% firms confirm that it is important, whereas 33% found it very important. The overall PPS score for this item was found 85.61%.

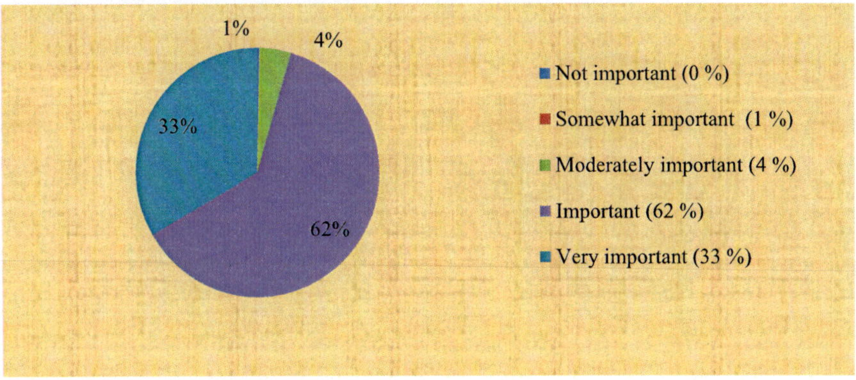

Fig. 8.5 Role of continuous improvement and innovation

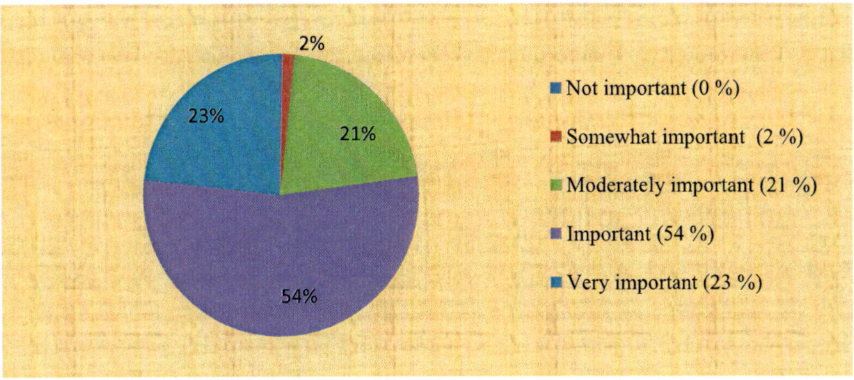

Fig. 8.6 Benchmarking

8.2.6 Benchmarking

The survey results on factor "Benchmarking," are presented in Fig. 8.6. It is observed that 54% respondents feel it is important, 23% very important, and 21% said it is moderately important. The overall PPS score for this item was found 68.67% respectively.

8.2.7 Technical Expertise

With respect to factor "technical expertise," the survey results are presented in Fig. 8.7. It shows the capabilities to perform the duties of one's profession generally or to perform a particular professional task, with skill of an acceptable quality,

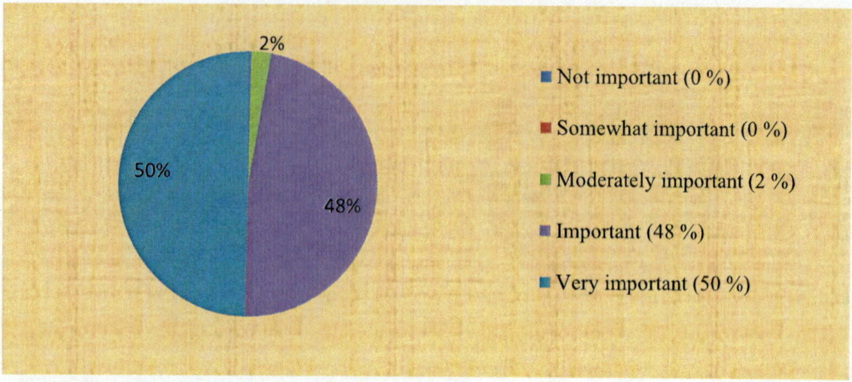

Fig. 8.7 Technical expertise

introduction of QM practices is rewarding. From figure, it can be seen that 50% respondents feel it very important, whereas 48% important. The overall PPS score for this item (89.23%) is fairly high.

8.2.8 Role of Market Expertise

With respect to subfactor "role of market expertise", survey results are presented with Fig. 8.8. From figure, it can be seen that 63% MSMEs feel it important, whereas 30% ticked on very important aspect. The overall PPS score for this item was found 84.45%.

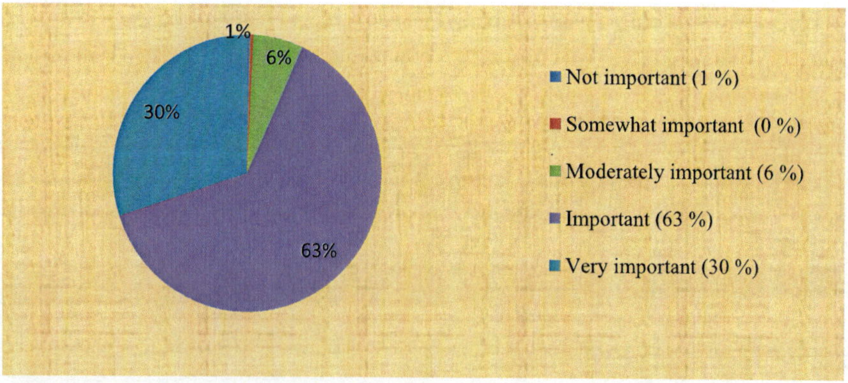

Fig. 8.8 Marketing expertise

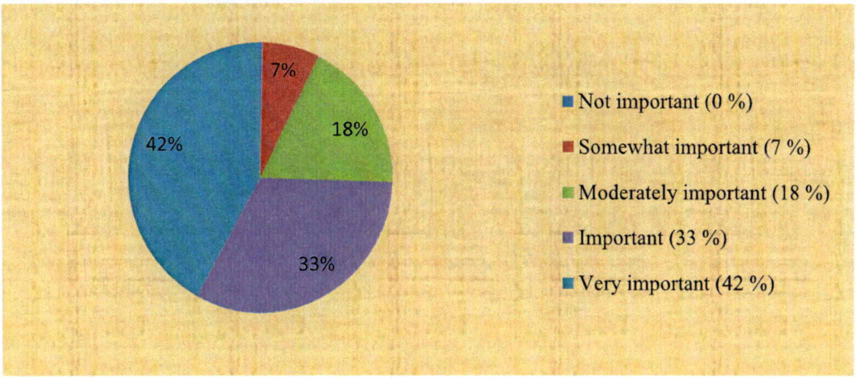

Fig. 8.9 Financial access

8.2.9 Role of Financial Access

With respect to last item in strategic factor "role of financial access," results were calculated and are presented in Fig. 8.9. Survey result found that 42% firms tick on very important, with 33% tick on important. The overall PPS score for this point was found very high, i.e., 93.92%, respectively.

8.3 Tactical Factors

With respect to "Tactical factors," there are total of seven subfactors that represent the importance of tactical factors.

8.3.1 Role of Employee Encouragement

To begin with the first item "role of employee encouragement," survey results so obtained are calculated and presented with the help of Fig. 8.10. It can be seen that 50% firms feel it very important aspect, whereas 49% feel it important. The overall PPS score for this item was found to be 89.44% respectively.

8.3.2 Role of Employee Involvement

The second subfactor "employee involvement" is quite relevant with first one, and the overall PPS point is also moreover same, i.e., 89.01%. This PPS score consists

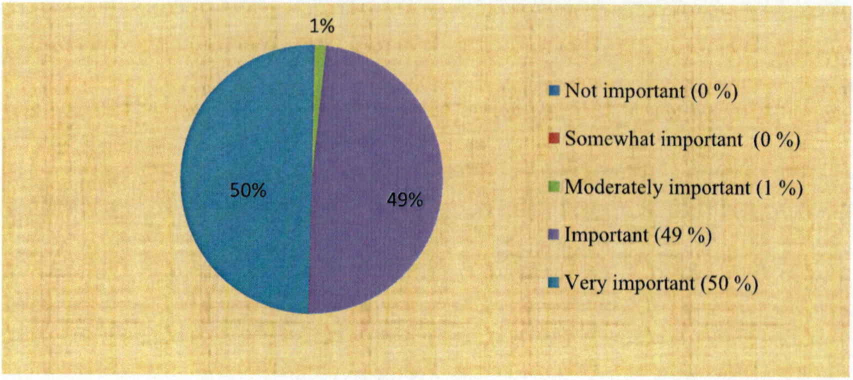

Fig. 8.10 Employee encouragement

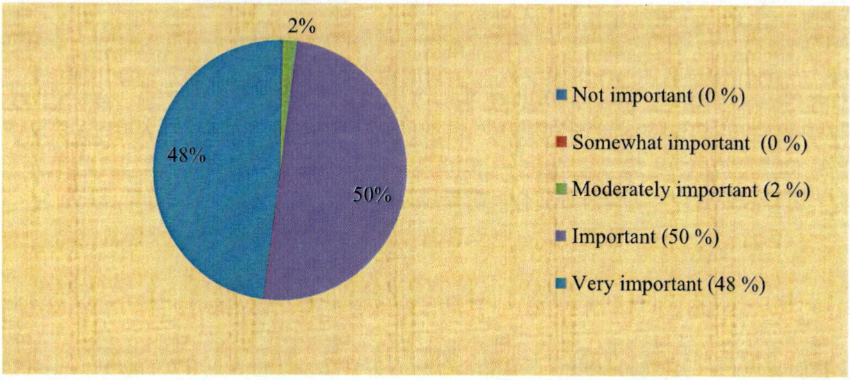

Fig. 8.11 Employee involvement

of responses with 48% very important and 50% important. Figure 8.11 shows the results in detail.

8.3.3 Training and Education

With respect to the third item, i.e., "training and education," survey results are calculated and presented in Fig. 8.12. It can be seen that 66% respondents said it is very important, whereas 32% said it an important aspect. The overall PPS score for this item was found very high 92.63%, which emphasize the role of training and education in successful introduction of QM.

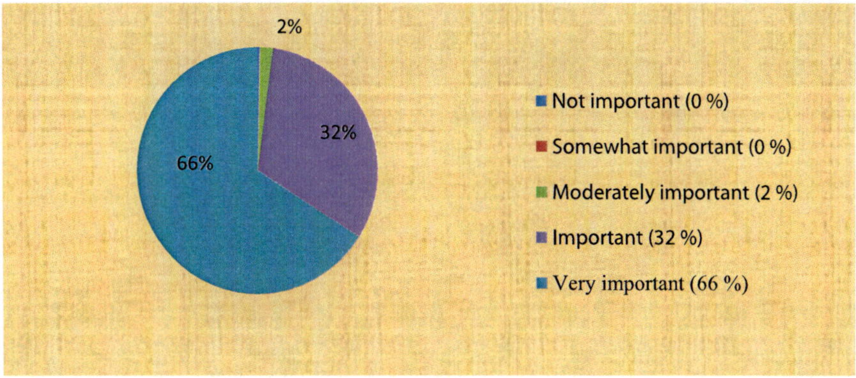

Fig. 8.12 Training and education

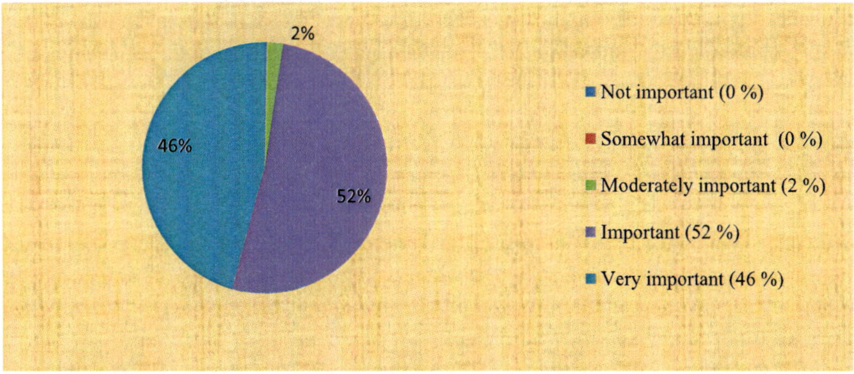

Fig. 8.13 Teamwork

8.3.4 Teamwork

With respect to fourth item "teamwork," it was found that 52% firms said it is important and 46% said it is very important aspect. The overall PPS point for this item was found 88.49% respectively as shown in Fig. 8.13.

8.3.5 Information and Analysis

Survey results with respect to "information and analysis" are presented in Fig. 8.14 which shows that 69% respondents feel it important, whereas 25% said it is very important. The overall PPS score for this item was found 83.87%, which is high as

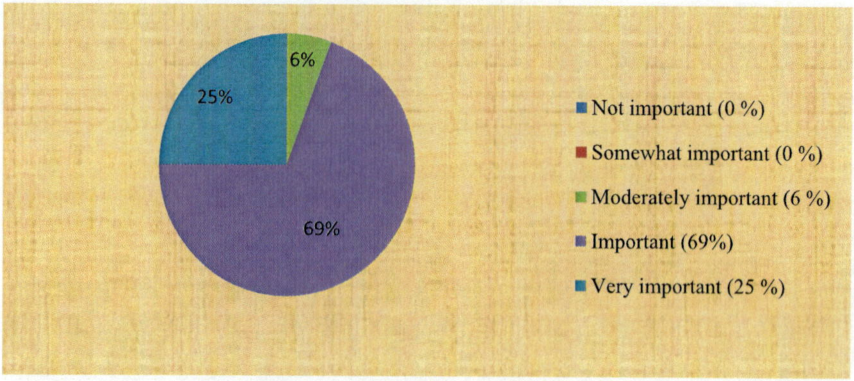

Fig. 8.14 Information and analysis

information processing and analysis is very important factor as it helps in forecasting in sales or predicting marketing behaviors for new or existing products.

8.3.6 Supplier Management

With respect to sixth item "supplier management," survey results are presented in Fig. 8.15. Results found that 66% firms feel it important, whereas, 18% feel it very important aspect in implementing quality management practices. The overall PPS score for this item was found 79.93%.

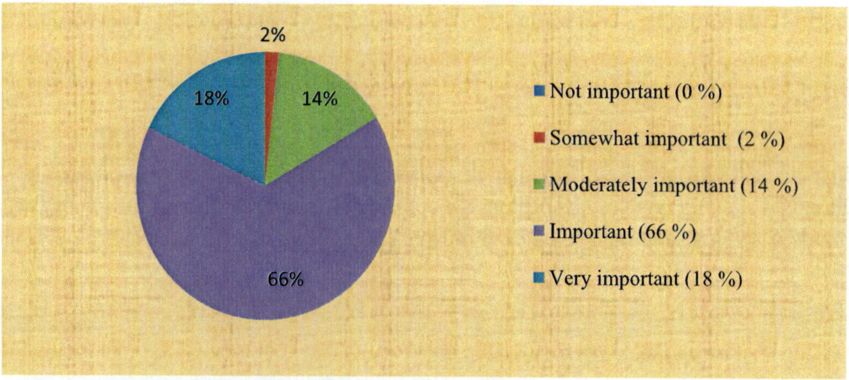

Fig. 8.15 Supplier management

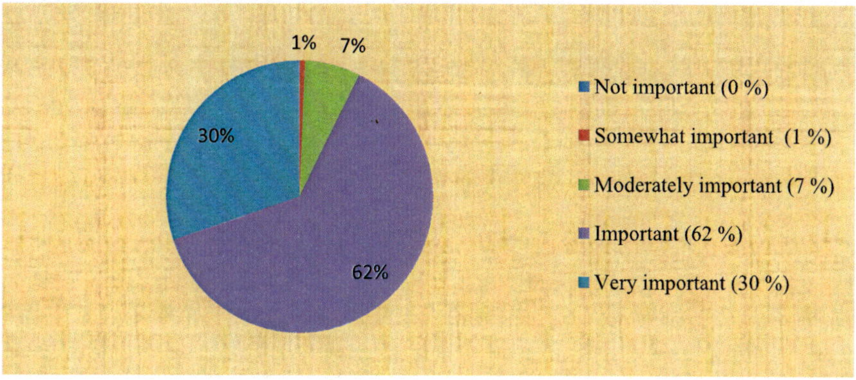

Fig. 8.16 Role of communication

8.3.7 Communication

With respect to last item in this dimension "communication," results are presented in Fig. 8.16. Survey result found that 62% respondents feel it important, whereas 32% found it very important aspect, in implementing QM practices. The overall PPS score for this item was found 84.42%.

8.4 Operational Factors

With respect to "Operational factors," there are total of six items that represent the importance of these factors.

8.4.1 Product and Service Design

For the first item "product and service design," survey results are presented in Fig. 8.17. It can be seen that 61% feel important, whereas 24% feel moderately important. The overall PPS score for this item is 77.46% respectively.

8.4.2 Process Management

For the second item "process management," survey results found that 58% MSMEs feel it important and 25% moderately important. The results with importance of process management are presented in Fig. 8.18. Overall PPS score for this item was found 77.68% respectively.

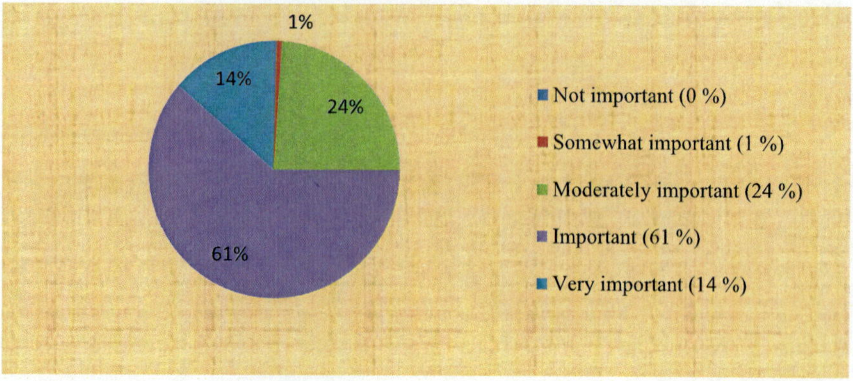

Fig. 8.17 Product and service design

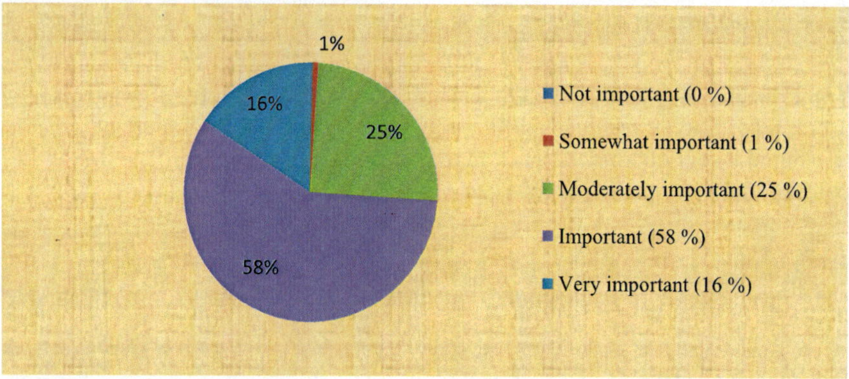

Fig. 8.18 Process management

8.4.3 Customer Focus

With respect to third item "customer focus," 68% respondents described it very important whereas 28% said it is an important aspect for successful implementation of quality management practices. Detailed result for this item is presented in Fig. 8.19. Overall PPS score for this item was found very high, i.e., 93.14%, which shows the importance of customer focus.

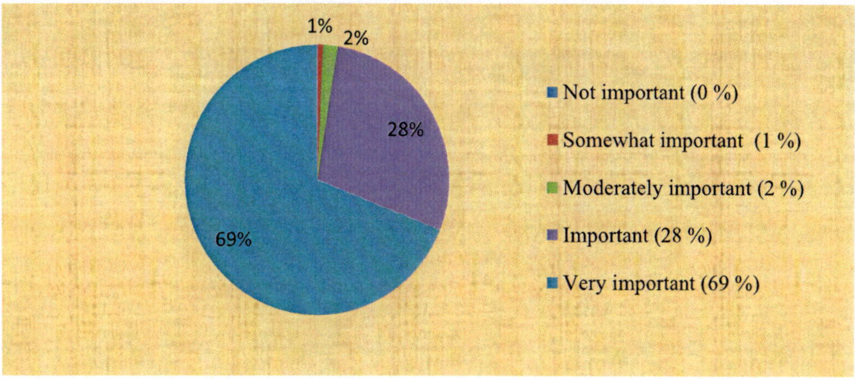

Fig. 8.19 Customer focus

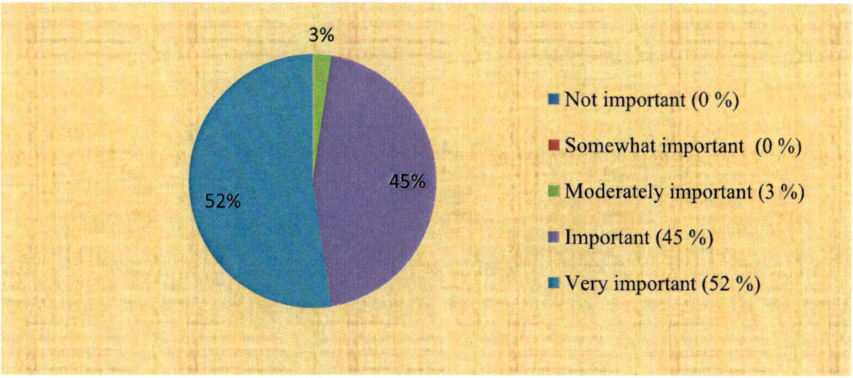

Fig. 8.20 Human resource management

8.4.4 Human Resource Management

With respect to fourth item "human resource management," survey results found that 52% MSMEs feel it very important and 45% MSMEs important. The detailed result for this item is presented with Fig. 8.20. Overall PPS score for this item was found 90.00%, which shows the importance of human resource management.

8.4.5 Industrial Relations

With respect to fifth item "industrial relations," results are presented in Fig. 8.21. It can be seen that 72% respondents ticked on important and 25% on very important. The overall PPS score for this item was found 84.14% respectively.

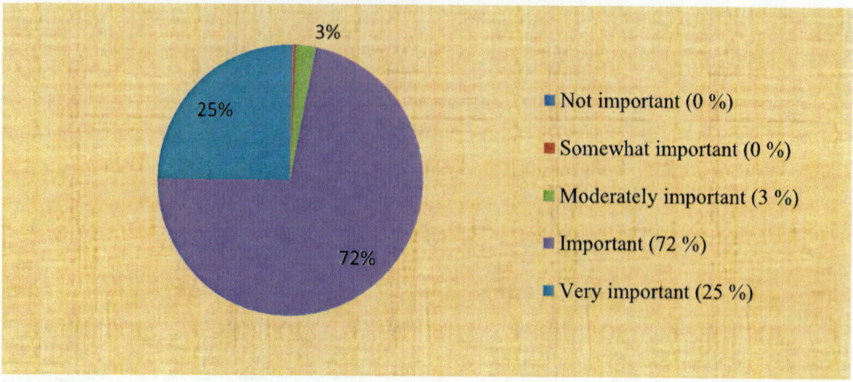

Fig. 8.21 Industrial relations

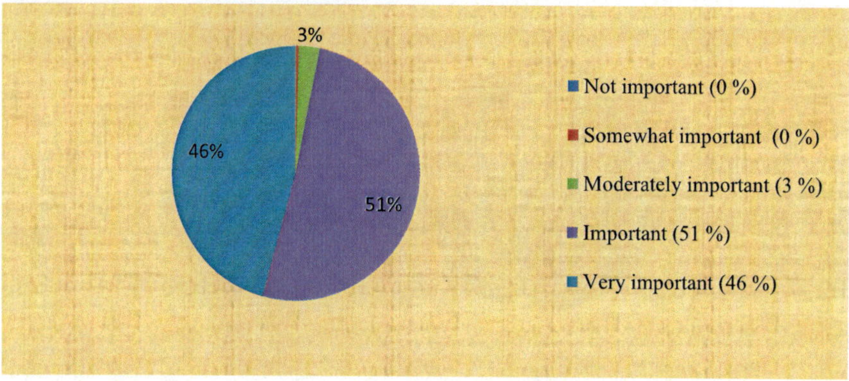

Fig. 8.22 Employee rewards and incentives

8.4.6 Employee Rewards and Incentives

The survey results for last item, i.e., "employee rewards and incentives," are presented in Fig. 8.22. It was found that 46% MSMEs said it is very important and 51% MSMEs said it is important. The overall PPS score for this item was found 88.46% respectively.

8.5 Concluding Remarks

This chapter presents the details with respect to various critical factors, which influence the implementations of quality practices in firms (e.g., Top management commitment, Quality culture, Quality system, Quality awards, Continuous

Improvement, Benchmarking, Employee involvement, Employee encouragement, Training and Education, Information and Analysis, Supplier Management, Communication system, Leadership quality, Product and service design, Process Management, Customer Focus, Human resource management, Industrial relations, Rewards and incentives). These factors are classified into three categories, i.e., (i) Strategic factors (ii) Tactical, and (iii) Operational factors. It is evident from the results that Top management commitment is one of the most important factors under strategic category with highest PPS score. In tactical category, Employee involvement and Employee encouragement share almost same highest PPS score. In operational category, the factor customer focus and human resource management have high PPS scores, i.e., above 90.0.

Chapter 9
Comparative Analysis of Competitive Positioning Among MSMEs Sectors

Highlights

- Identified the issues contributing to firm's competitiveness.
- Developed a comparative case study by calculating the overall surface areas of firms operating in various key MSME sectors based upon Porter's diamond framework.

9.1 Introduction

In modern global market due to globalization and rapid technology changes, firms need to compete not only with national rivals but also with international companies (Ickis 2006). This immense global pressure continues to alter the environment in which firms operate; the traditional industrial strategies are becoming less effective. For survival, companies have to build "core competencies" via implementing quality practices, cost-effective competitive pricing policy, internet marketing, sound strategy basis, product innovation and predicting buyer behavior for high customer satisfaction (Chobanyan and Leigh 2006). To this effect, the present chapter presents a case of MSMEs situated in the state of Himachal Pradesh, and the most popular approach to measuring competitiveness among various sectors, the framework developed by Michael Porter has been used.

To determine the competitive advantage of various MSME sectors, i.e., pharmaceutical, electronics, automobile, food and textile sectors, the individual diamond's axes dimensions related to various determinants (factor conditions, demand conditions, related and supporting industries, firm's strategy, structure and rivalry and Government and Culture) of porter model were determined.

R .K. Sharma, *Quality Management Practices in MSME Sectors*,
Springer Tracts in Mechanical Engineering,
https://doi.org/10.1007/978-981-15-9512-7_9

9.2 Michael Porter's Competitive Approach

Michael E. Porter has been regarded as a leading authority on competitive strategy in a nation and more recently the application of competitive analysis on social and environment aspect of business activities. According to Porter's (1998) competitive, advantage of the nation is determined by the strength of their factor endowments, their demand conditions, and the competitiveness of firm strategy, structure, and rivalries in the leading industries. Figure 9.1 shows the Porter's Diamond model of national competitive advantage.

9.2.1 Factor Conditions

These conditions contribute to the productivity of an organization, act as the key determinants of the level of prosperity of an organization can sustain over time. Factor conditions have been highlighted in recent literature: based on basic factors or tangible resources such as climate, location, available minerals, national resources,

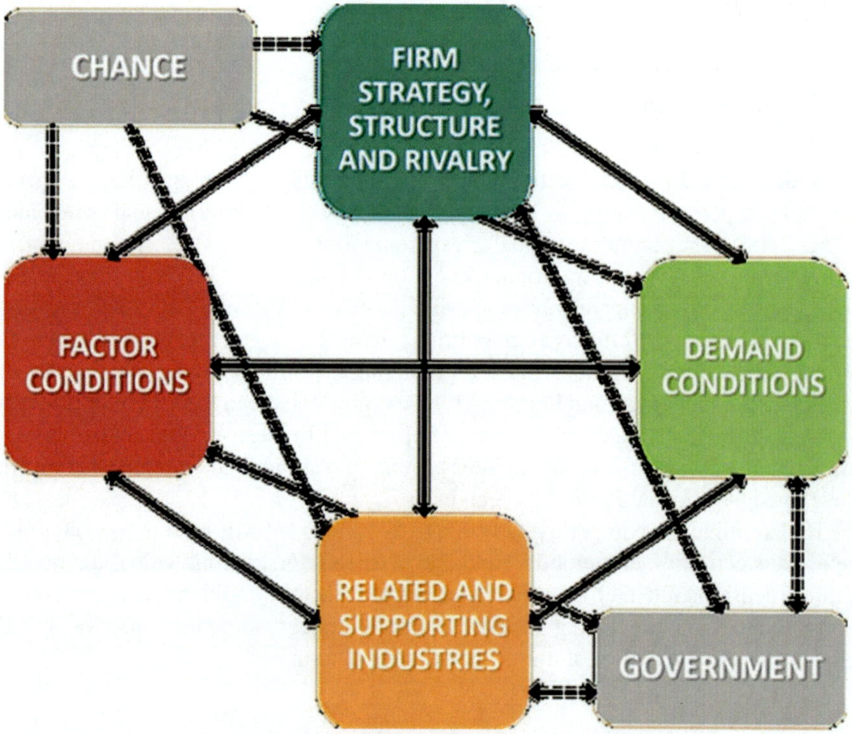

Fig. 9.1 Porter's diamond framework

agriculture advancements, forest resources, skilled and unskilled labor, and advance factors or intangible resources (Porter 1998; Denizet al. 2013). The second category consists of advance factors, which are mainly affected by human efforts or are created to generate competitive advantages. For an instant, the competitive human resource includes abilities, skills and satisfaction level of employees (Nanda and Singh 2009). The comapritively low cost of staff helps in national competitivness and economic growth (Hamalainen 2003). Physical resources include raw materials, its quality and quantity, efficient utilizations of limited available resources, knowledge resource consist of the state's stock of scientific, technical and market knowledge (Petrakis et al. 2015). Information resources include universities, government research institutes, government statistical agencies, business and scientific literature, reports and database, etc. (Nair and Ghosh 2006).

9.2.2 Demand Conditions

This determinant refers to the nature of home market and is the second broad determinant of national competitive advantage. Porter viewed demand conditions regarding the size of the home market, sophistication and demanding buyers (either industrial buyers or consumers). Traditionally, the demand conditions were thought of as the size of a local market. Porter adds another perspective by focusing explicitly on the quality of domestic demand: specific customer's need at a given location can provide companies with the unique ability to learn how to serve these requirements with targeted products and services. If local customers need to foreshadow the deeds of clients in the other markets, firms that operated locally are likely to have a competitive edge when they enter the new market. Furthermore, he emphasized that the focus of local demand not imply that global market access does not matter (Porter 2004; Delgado et al. 2014).

9.2.3 Related and Supporting Industries

The geographic concentration of companies and other institutions active in a particular economic field are essential for competitiveness because they are associated with higher level of productivity and innovation. Related industries are those in which an organization can organize or allocate activities in the value chain when competing or those, which produce basic goods. The existence of related or supporting industries in a nation is argued as the third dimension of the diamond model. According to Nair and Ghosh (2006), the existence of competitive downstream industries well contributes to the four traditionally primary applied competitive priority dimensions, i.e., (i) cost, (ii) flexibility, (iii) quality and (iv) delivery. Furthermore, the presence of suppliers and related industries within a nation that are internationally competitive provides benefits such as innovation, upgrading of information flow and shared technology development through firm alliance to create an advantage in downstream industries. They provide access to specific input factors such as employees and risk capital to enable higher firm's performance. Potential suppliers not only reduce the delivery lead time but also invest in research, innovative activities and create knowledge base as per upstream enterprises.

9.2.4 *Firm Strategy, Structure and Rivalry*

This determinant refers to the conditions, which determine how the firms were created, organized, structured and managed, as well as the nature of the domestic rivalry. The feature that governs this determinant is unique to countries vary as to the lifestyle of people, their attitude individual as well as the group toward authority, which further constitute into organizational culture (Cho et al. 2013; Esen and Uyar 2012). The educational system, social and religious belief, history, and family structure directly affect the structure of the business. This determinant governs the uniqueness of company over the competitor. This determinant governs the uniqueness of company over the competitor. In other words, it shows the firm's competitive priority such as increased productivity using new equipment, or unique product and production capabilities (Lin and Wu 2014). Effective production planning system and implementation of quality improvement techniques investment in R&D activities, technology sharing and upgradations (Nanda and Singh 2009), or using strategic management and skills for integration of business activities. With respect to this determinant, Porter's concept focuses on creating prosperity by achieving a market price for their output more than the cost of providing this output (Porter et al. 2004). According to him, the changes in economic policy will only affect prosperity if organizations make changes in the nature and extent of activity undertaken by companies. On the other hand, the pressure from local rivals makes intimate access to new ideas and lower the cost associated with forming new products and cultivates higher productivity growth.

9.2.5 *Government and Chance Events*

Governments establish an environment that fosters a high standard of living for its citizens by making policies regarding health, safety, environment, financial structure, industrial trade and development (Brown et al. 2010; Doglet al. 2012; Denizet al. 2013). The government consists of many distinct agencies and geographic units with their impact on the business environment at a given location. Government policies directly or indirectly affect the competitive environment for enterprises, which may further yield into minimizing the external risk for economic activities. According to Kang and Park (2012), a well-developed government infrastructure supports economic activities—it includes access to natural resources, functional business systems, information technology, transportation and communication systems, educational structure and protection of the environment. The role of chance is often included in the Diamond Model as the likelihood that external events such as war and natural disasters can negatively affect or benefit a country or industry. However, it also includes random events such as where and when fundamental scientific breakthroughs occur. These events are beyond the control of the government or individual companies. For instance, the heightened border security, resulting from the September 11 terrorist attacks on the USA undermined import traffic volumes from Mexico, which has had a large impact on Mexican exporters. The discontinuities created by chance may lead to advantages for some and disadvantages for other companies.

9.3 Current Competitive Positioning of MSME Sectors

Diamond Axes Length

To determine the competitive advantage of various industrial sectors, we calculated separate diamond and compared them as per studies proposed by Rahman (2001), Herciu (2013), and Dogl et al. (2012). Within the framework of Diamond Model, the competitiveness of the firms was measured with the scale to the scores obtained by various determinants. Such as factor conditions (two casual variables, nine proxy variables); demand conditions (two casual variables, nine proxy variables); related and supporting industries (two casual variables, eight proxy variables); firm strategy, structure and rivalry (two casual variables, eight proxy variables); government and culture (two casual variables, nine proxy variables) as shown in Table 9.1. For the measurement of proxy variables, an interval scale with a minimum of 0 and a maximum of 10 is computed. If a casual variable is determined by more than one proxy variables, the arithmetic average is calculated. The results are obtained in one score with values between 0 and 10.

Thereby, each determinant was determined by a composite score of two causal variables, which were itemized by different proxy variables for the competitive advantage of industry. For the purpose of construction and interpreting the diamond about the size of the axes and the surface area, we added the two causal variables. Afterward, the diamond surface areas for different sectors were calculated and analyzed. Table 9.1 shows the details with respect to the determinants and variables. The descriptive results with respect to scores under the MSME sectors are presented in Table 9.2 and Fig. 9.2a–e presents the graphical representation about the findings related to diamond axes.

Table 9.1 Determinants of Porter's diamond

S. no	Determinants	Casual variable	Interval scale	Proxy variable	Number
1	Factor conditions (max. 10)	Basic	(1–10)	F1, F2, F3, F4, F5, F6, F7, F8, F9	9
		Advance			
2	Demand conditions (max. 10)	Market volume	(1–10)	D1, D2, D3, D4, D5, D6, D7, D8, D9	9
		Sophistication			
3	Related and supporting industries (max. 10)	Related companies	(1–10)	RS1, RS2, RS3, RS4, RS5, RS6, RS7, RS8	8
		Support			
4	Firm structure strategy, and rivalry (max. 10)	Rivalry	(1–10)	FS1, FS2, FS3, FS4, FS5, FS6, FS7, FS8	8
		Structure/strategy			
5	Government and culture (max. 4)	Government support and culture	(−4–4)	GC1, GC2, GC3, GC4, GC5, GC6, GC7, GC8, GC9	9

Where F1: Natural resources, F2: Physical resources, F3: Unskilled labor, F4: Skilled labor, F5: Production and process technology, F6: Scientific and Technological Information, F7: Capacity Utilization, F8: Communication infrastructure, F9: IPR and paten, D1: Size of domestic demand, D2: Pattern of growth, D3: Export potential, D4: Periodically Increase in demand, D5: Brand value, D6: New investment in the region, D7: Distribution channel, D8: Bureaucracy and control, D9: Service efficiency level, RS1: Level of technology upgradation, RS2: Quick information flow, RS3: Shared technology development, RS4: Level of active work of relevant civil societies, RS5: R & D investment, RS6: Suppliers and distribution channel, RS7: Marketing Support, RS8: Relations with research and development institutes, FS1: Boost innovation, FS2: Level of marketing innovation, FS3: Level of quality management practices, FS4: Information flow with firm, FS5: Quality awards, FS6: Internal structure, FS7: Geographic concentration, FS8: Quality of supplier, GC1: Financial support system, GC2: Environmental regulation, GC3: R & D support, GC4: Training facilities, GC5: Industrial land allocation, GC6: Labor Laws, GC7: Impact of national culture, GC8: Business climate, GC9: Formal and informal rules

9.3.1 Pharmaceutical Sector

With emerging competitiveness and entrepreneurial skills, this MSME sector is conspicuous by the large presence of private sector and has captured a substantial share of the domestic and external market. The sector's score in factor conditions is 8.15, which indicates its strength, especially in advance factor conditions such as high technology types of equipment, skilled manpower and knowledge resource. This sector produces a wide range of drugs to meet high demands, from different segments of consumers, as revealed by its high score (8.78) in demand conditions. Further, very high score in third and fourth determinants 9.13 and 8.17, respectively, indicates high competitive advantage from supporting industries, strategies of distributions of resources in downstream firms.

Table 9.2 Sector-wise score of key MSMEs sectors

Sr. no	Determinants	Casual variables	Pharmaceutical	Electronics	Automobile	Food	Textile
1	Factor conditions (max. 10)	Basic	7.75	5.03	7.69	8.69	5.22
		Advance	8.56	7.04	6.21	6.21	4 20
		Sum	**8.15**	**6.04**	**6.95**	**7.45**	**4.71**
2	Demand conditions (max. 10)	Market value	8.79	8.90	6.34	8.79	7.12
		Sophistication	8.76	7.65	7.04	7.82	8.25
		Sum	**8.78**	**8.28**	**6.69**	**8.30**	**7.69**
3	Related and supporting industries (max. 10)	Related companies	9.57	6.82	6.22	7.69	4.24
		Support	8.68	5.14	7.18	8.53	4.32
		Sum	**9.13**	**5.98**	**6.70**	**8.11**	**4.28**
4	Firm strategy, structure and rivalry (max. 10)	Rivalry	8.67	7.28	7.57	6.05	6.65
		Structure/strategy	8.74	6.36	6.23	6.29	6.54
		Sum	**8.71**	**6.82**	**6.90**	**6.17**	**6.69**

(a)

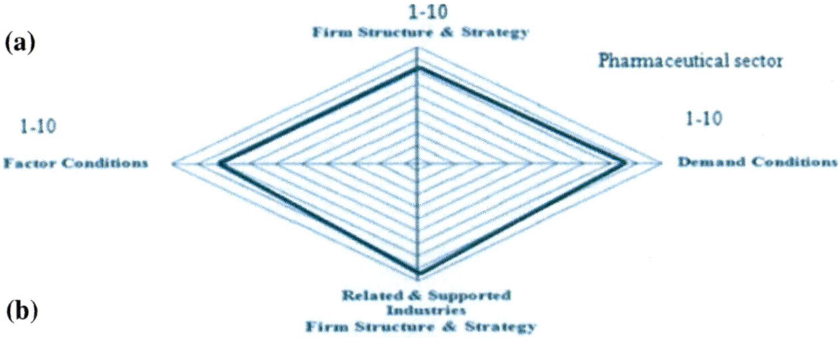

(b)

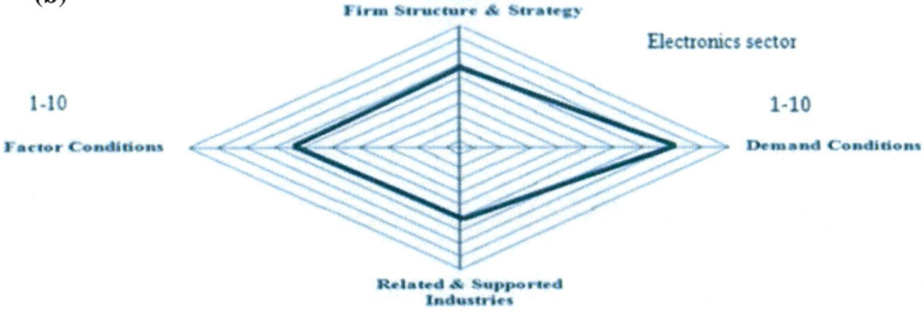

Fig. 9.2 a–e Graphical representation of diamond axes of MSME sectors

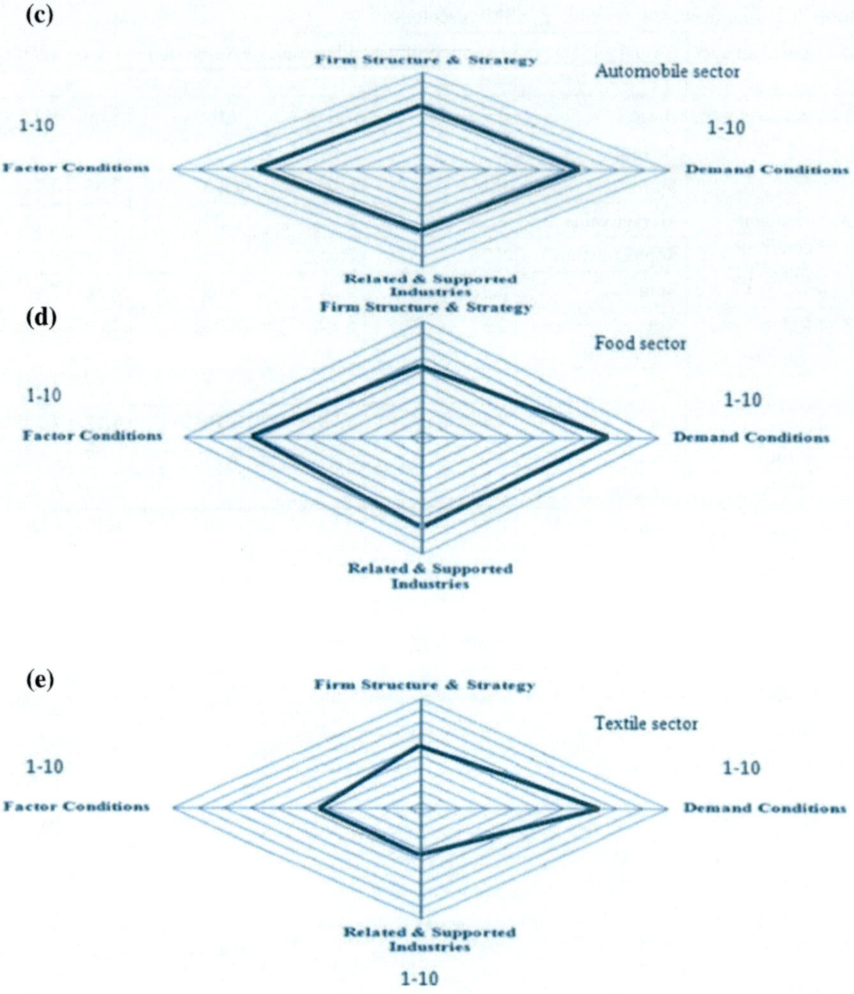

Fig. 9.2 (continued)

9.3.2 Electronics Sector

The high value of second determinant (8.28) shows the requirement of electronics goods in the Indian market. The average scores in first, third and fourth determinants 6.04, 5.98, and 6.82, respectively, show challenges with respect to manpower with required skills, overall supports from related industries and understanding of firms regarding the importance of strategy making in perspective of current market scenario.

Table 9.3 Diamond surface areas for MSME sectors

Diamond surface area	P	E	A	F	T
ASD = firm strategy, structure and rivalry * ½ demand conditions	38.25	28.23	23.08	30.91	25.72
ARD = related and supporting industries * ½ demand condition	40.14	24.75	22.41	33.66	16.46
ARF = related and supporting industries * ½ factor condition	37.20	18.06	23.28	32.21	10.08
ASF = firm strategy and rivalry conditions * ½ factor conditions	38.25	20.65	23.98	22.98	15.75
Sum (units)	**153.85**	**91.79**	**92.75**	**119.76**	**68.01**

Where: P = Pharmaceutical, E = Electronics. A = Automobile, F = Food, T = Textile

9.3.3 Automobile Sector

Practicing of quality management philosophies such as, 5S and Housekeeping, TPM, etc., in the automobile sector have provided boost to improve quality in this sector as evident from the moderate score in all four determinants as given in Table 9.1 and Fig. 9.2c. Availability of required skilled and unskilled manpower (6.95) is one of the important variables under factor conditions, which need to be addressed to improve competitiveness. Further, as India being a developing country demand for automotive is constantly increasing (6.69). Strong supply chain network and firm strategy to meet upcoming demands were found appreciable as indicated by medium score 6.70 and 6.90, respectively.

9.3.4 Food Sector

After pharmaceutical, the food sector shows high growth in the region. This sector establishes a vital linkage and synergy between the two pillars of the Indian economy (i) industry and (ii) agriculture. The high demand condition (8.30) makes India world's second largest producer of food. The enormous growth potential of this sector can be understood from the fact that food production in the country is expected to double in the next 10 years while the consumption of value-added food products will also correspondingly grow. The high value of related and supported industries (8.11) shows the raising agriculture yield, enhancing productivity, creating employment and raising life standard of a large number of people across the country, especially those in rural areas. Furthermore, Fig. 9.2d shows the high (7.45) and moderate (6.17) score obtained by the sector in remaining two determinants (i) factor conditions and (ii) firm structure and strategy, which show the ability of this sector to sustain high competitive advantage for a long period of time.

9.3.5 *Textile Sector*

This sector is facing competition in productivity, delivery schedule, the reliability of products and other such intangible factors like the image of the country/company and brand quality. This sector shows medium score in factors condition and support from related industries, i.e., 4.71 and 4.28, respectively. Whereas, owing to high demand conditions (7.69), this sector shows tremendous scope for new technology, brands, and quality products through new competitive strategy development. It can be seen in Fig. 9.2e, which shows comparatively smaller diamond for this sector.

Diamond Surface Areas

After reporting the main findings for each dimension of the diamond separately, the diamond surface areas for all industrial sectors considered in the study are calculated by summing up the individual areas of each quadrant's triangle as shown in Table 9.3.

Porter suggested that a large diamond represents high competitiveness and a small diamond represents low competitiveness (Dogl et al. 2012). The overall surface area of pharmaceutical sector is very high as 153.85, it shows its high competitive positioning among other sectors as represented in Fig. 9.3. The high growth of the pharmaceutical sector has attracted global players to India and leaders like Ranbaxy Laboratories Ltd., Dr. Reddy's, Sun pharmaceutical, Cipla Ltd. and many more have made large investments to access the Indian market.

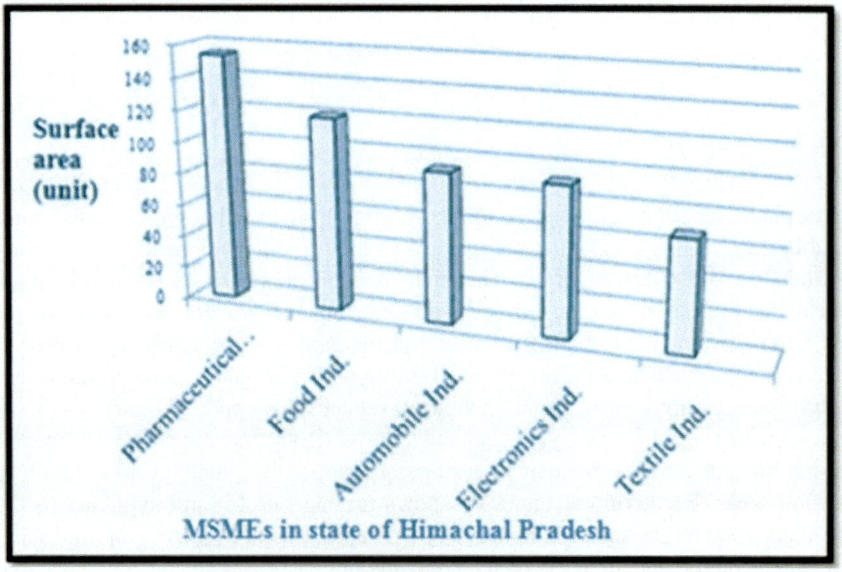

Fig. 9.3 Sector-wise surface area/competitiveness positioning of MSMEs

The overall surface area of electronics and automobile sector is found moderate, i.e., 91.79 and 92.75, respectively. This is because of the consequent effect of low score in third determinants of diamond model. The high surface area in food sector, i.e., 119.76 shows potential scope of food technology, investment in new project and environment for quicker development in this sector.

The food industry has started producing many new items like ready-to-eat food, beverages, processed and frozen fruits and vegetable products, marine and meat products, etc.

The Indian consumer is being fast introduced to newer high-quality food products made by using the latest state-of-the-art technology, which is also giving the industry a competitive edge. One big threat to this sector would be China because, from last few decades, China has upgraded its technology faster than India, and exporting appreciable higher good as compared to India. The low score of Porter's model surface area (68.01) in textile sector shows that this sector required structural reforms. Indian textile and clothing sector have a tremendous potential, only a portion of which has been exploited due to policy constraints.

However, there lies a considerable potential that has not been exploited primarily due to government policies, limited financial resources and lack of entrepreneurship skills in the region. Since garment manufacturing is reserved for small-scale industries in India, catering to the small size they cannot create competitive products. In India, most of the textile mills are organized one, but the demand for Indian garments overseas is fashion driven.

9.4 Concluding Remarks

The results presented in this section indicate that, about 99% of firms operating in selected industrial clusters are quality conscious and about 70% of them are ISO certified and the main reasons for introducing QM practices by them are to increase productivity, earn profitability and sustain long-term growth and dynamism. Most of the firms have adopted Quality Practices/Programs since 2005 and almost all are using QM practices in all stages of work, i.e., inspecting incoming raw materials, work in process and final/finish products. Gaining quality standard certification helps the firms to become competitive in domestic as well as in foreign market.

The results with respect to the application of Porter Diamond model for competitiveness indicate that the development of industrial clusters helps to strengthen innovativeness and the promotion of cooperation with local/foreign partners in R&D investment and shared technology development/upgradation. From the analysis of results, it is observed that competitiveness among MSME sectors is mostly affected by intangible resources, which are difficult to imitate by competitors and by the demand conditions followed by firm strategy, structure and rivalry and, supportive industries. Moreover, the diamond axes indicate significant difference with respect to determinants, within various MSME sectors in the region which can help the managers to perform SWOT analysis and focus on those areas where they have

Table 9.4 Summarized views of Porter's diamond of the MSMEs

Diamond determinants	Competitive advantage	Competitive disadvantage	Opportunities
Factor conditions	• Skilled human resource • Availability of local labor • Production and process technology • Low initial investment • Investor protection	• Availability of financial service • Quality of overall infrastructure • Contractual labor • Lack of funding for in-house research initiative	• Employment opportunities for weaker section of society • Training opportunities at government institute and big firms
Demand conditions	• Market size • Different segments of buyer • Number of procedure to start a business	• Degree of customer orientation • Poor investment in R&D • Lack of marketing knowledge	• Increasing demand growth in each sector • Increase curiosity towards new products
Related and supporting industries	• Presence of international competitive front-end industries that effectively coordinate global supply chain management • Effective communication • Rapid movement of ideas	• Local supplier quality • Local supplier quantity • State of cluster development	• Have greater learning opportunities by collaborating with large scale units and institutes like IITs, IIMs NITs,etc
Firm strategy, structure and rivalry	• Functional and flexible management • High level of commitment • High ownership feeling • Quality of math and science educations • High specific return	• Firm-level technology absorption • Capacity for innovation • Little career planning for employee • Little formal training program	• Participation by entrepreneurs in international trade can lead to diversification of markets • Implementation of advanced quality tools cum techniques

(continued)

Table 9.4 (continued)

Diamond determinants	Competitive advantage	Competitive disadvantage	Opportunities
Government and culture	• Central excises exception • General government debt	• The bureaucratic hurdle in obtaining govt. support • Not properly define financial strategy and corruption	• Transparency in public policy, Single window Clarence system, Tax benefits, reservation of production for MSMEs

strengths and to introduce corrective measures to overcome potential weaknesses. For instance, in the textile sector, the determinants such as factor conditions and related and supported industries scored 4.71 and 4.28, respectively, which needs to be strengthened as this sector stands at last position with a minimum surface area of 68. Whereas, owing to high demand conditions (7.69), this sector shows tremendous scope for new technology, brands, and quality products through new competitive strategy development. Table 9.4 presents the summarized views of competitive advantage, competitive disadvantage and opportunities related to development of MSMEs.

References

Brown P, McNaughton RB, Bell J (2010) Marketing externalities in industrial clusters: a literature review and evidence from a Christchurch, New Zealand electronics cluster. J Int Entrepreneurship 8(2):168–181

Cho Y-H, Yu G-C, Joo M-K, Rowley C (2013) Changing corporate culture over time in South Korea. Asia Pac Bus Rev 20(1):9–17

Chobanyan A, Leigh L (2006) The competitive advantages of nations: applying the diamond model to Armenia. Int J Emerg Mark 1(2):147–164

Delgado M, Porter ME, Stern S (2014) Clusters, convergence, and economic performance. Res Policy 43(10):1785–1799

Deniz M, Seckin SN, Cureoglu M (2013) Micro-economic competitiveness: a research on manufacturing firms operating in TRB1 region. Procedia-Soc Behav Sci 75:465–472

Dogl C, Holbriugge D, Schuster T (2012) Competitive advantage of German renewable energy firms in India and China: an empirical study based on Porter's diamond. Int J Emerg Mark 7(2):191–214

Esen S, Uyar H (2012) Examination of competitive structure of Turkish tourism industries in comparison with diamond model. Procedia: Soc Behav Sci 62:620–627

Hamalainen TJ (2003) National competitiveness and economic growth. Edward Elgar Publishing Ltd., Gloucestershire

Ickis JC (2006) Building a national competitiveness program. J Bus Res 59(3):341–348

Kang KN, Park H (2012) Influence of government R&D support and inter-firm collaborations on innovation in Korean biotechnology SMEs. Technovation 32(1):68–78

Lin Y, Wu L (2014) Exploring the role of dynamic capabilities in firms performance under the resource-based view framework. J Bus Res 67(3):407–413

Nair SK, Ghosh S (2006) Managerial work values in India: a comparison among four industry sectors. South Asian J Manag 13(3):45

Nanda T, Singh TP (2009) An assessment of the technology innovation initiatives in the Indian small manufacturing industry. Int J Technol 9(2):173–207

Petrakis EP, Kositis CP, Valsamis GD (2015) Innovation and competitiveness: culture as a long term strategy instrument during the European great recession. J Bus Res 66(7):1436–1436

Porter M (1998) The competitive advantage of nations. Free Press, New York

Porter ME (2004) Building the microecomoic foundation of prosperity: findings from the business competitiveness index. World Competitiveness Report, 2005

Appendix A
Case Study 1

 # Integrating Six Sigma Culture and TPM Framework to Improve Manufacturing Performance in SMEs

Summary

Worldwide, small and medium enterprises (SMEs) have been accepted as the engine of economic growth, especially in the emerging economies, and for promoting equitable development their contribution as a component to the nation's infrastructure is relatively high.[1] Global competition forces SMEs to improve their competitiveness by enhancing their manufacturing performance. Hence, they have to pay attention to the reliability of their production processes as well as their commitment to quality management practices. In present times, because of automation and large-scale mechanization, higher plant availability, better product quality and longer equipment life had assumed considerable significance among the SMEs. In order to meet the above challenges, adoption and practice of proactive maintenance strategies along with execution of best management practices, i.e., lean Six Sigma (LSS), have become essential for organizations to bring down maintenance and production costs. As evident from literature, both total productive maintenance (TPM) and Six Sigma are key business process strategies employed by SMEs to enhance their manufacturing performance. However, whilst there is significant research information available on implementing these systems in an independent manner, there is little information available relating to the integration of these approaches to provide a unified and highly effective implementation framework? The purpose of this case study is thus to develop an integrated model based on Six Sigma and TPM framework focusing

[1] https://onlinelibrary.wiley.com/doi/abs/10.1002/qre.1525.

R .K. Sharma, *Quality Management Practices in MSME Sectors*,
Springer Tracts in Mechanical Engineering,
https://doi.org/10.1007/978-981-15-9512-7

on adding performance indicators. The goals of the study were achieved by utilizing various tools such as brainstorming sessions, pareto analysis fish-bone diagrams, histograms, FMEA, box plots, control charts and process capability plots for analysis. While implementation of such twin framework requires greater management commitment in terms of training, resources and integration, they are also expected to provide higher levels of equipment and plant performance as evident from the results. The results show significant improvement in OEE level (before 50, 54 and after 76, 83 for M/C-I and M/C-II), reduction in rework (from 22 to 10%), reduction in maintenance versus operation cost (from 30 to 10%) and reduction in defect rate (from 24.82 to 5%) and average outgoing quality level (from 30 to 5%). The substantial improvement in sigma level from 2.16 and 2.64 to 4.01 and 4.12 for M/C-I and M/C-II resulted in financial savings of approximately $2 m per annum.

 # Sharma R.K, Sharma R, Quality Reliability Engineering Int. 2014, Vol. 30 pp. 745–765.

Illustrative Case

The model framework (Fig. A.1) has been applied in a semiautomated paper manufacturing cell consisting of two paper machines, PM-I and PM-II. For the production of the paper, the raw material (soft + hardwood and bamboo) is chopped into small pieces of approximately uniform in size and transported to the store for temporarily

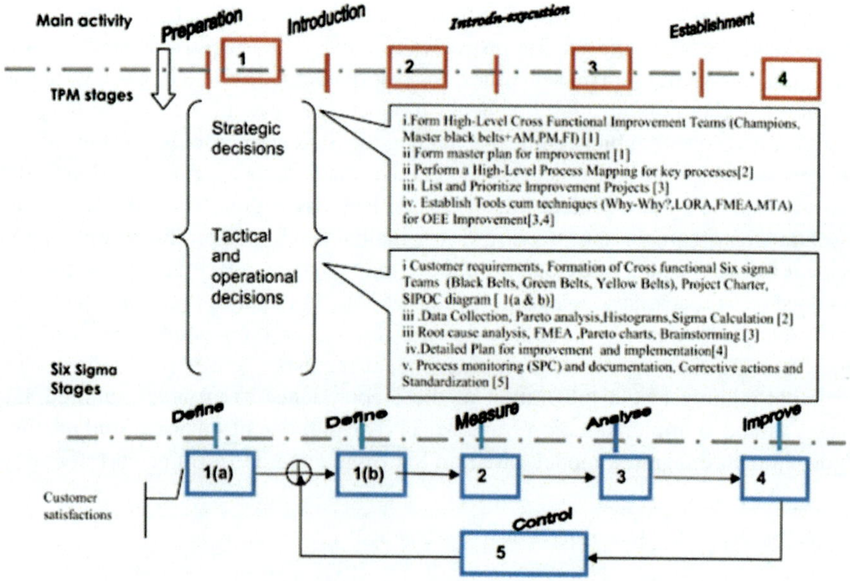

Fig. A.1. TPM and Six Sigma framework

storage by the use of a compressed air. A chain conveyor carries the chips from the store to digesters whenever required, where these are cooked using NaOH + Na_2S and steam at 8.5 kg/cm^2 pressure and around 180 °C. The cooked chips are called pulp. The pulp is transported to the storage tanks from where it is further processed through fibrelizier and refiner. The pulp is then filtered through filters and is washed (in three–four stages) with water to remove knots and chemicals. The washed pulp obtained in last stage of washing is stored in a surge tank. The next stages of processing are bleaching and screening. For the production of white paper, pulp is bleached (by passing chlorine gas through the pulp stored in the tank) and the brown pulp (used for packaging purpose) is screened directly. This pulp is then washed, cleaned and is sent to the headbox of paper machine consisting of three sections, i.e., forming, press and dryer, respectively.

On the quality front, Paper machine, PM-I and Paper machine, PM-II were working on 2.56 and 2.84 sigma levels, respectively, which were quite low.

Stage 1—Preparatory (Define Phase)

For successful implementation of TPM and Six Sigma culture in the cell, TPM office commonly known as TPM secretariat headed by senior executive of the company was formed. The main task of the office was to frame policies/set targets and to coordinate the activities for successful implementation and promotion of TPM and Six Sigma culture. To assist operators in performing maintenance tasks, cross-functional teams with members from (AM), (PM) and (FI) teams headed by respective group leaders, with members from engineering, maintenance group and production were formed. The team selected for implementation of DMAIC methodology in the cell includes the Senior Manager–Production as the MBB. The other members of the team were managers from planning, procurement, and quality and maintenance sections. The primary responsibility of team members was to support MBB in executing the project-related actions. The Head of Production department was identified as the Champion.

To create a common vision and sense of ownership for the project, the team members along with the Champion and Black belt developed a project charter mentioning all necessary details such as scope and schedule of different activities of the project. Black belt individuals have undergone extensive training in the use of technical tools and are responsible for carrying out the implementation of Six Sigma. They are entrusted with the task of measuring, analyzing, process controlling and coordinating numerous improving activities by acting as coaches, team leaders and facilitators in continuous improvement (CI) process. The team members after having discussions with the Champion decided to consider the rejection percentage of paper as the Critical to Quality characteristic for this project with goal to bring down the rejection level. Before implementation of the proposed framework, all the present problems related to maintenance and quality aspects were reviewed and are summarized in brainstorming session (shown in Table A.1).

Table A.1 Review of present problems

Autonomous maintenance AM	• Minor stoppages/repairs need technician intervention • Operators were not technically skilled to perform AM tasks. It requires the skill, adequate knowledge regarding the functional and failure aspects of the components. (Hence, skill improvement was realized essential) • No 5S activities in workplace (need better-disciplined work procedures at place) • As usual waiting for technician to do even minor maintenance work (I operate you fix, syndrome) • Lack of employee morale and attitude (need reward and recognition)
Planned maintenance PM	• Breakdown maintenance was in practice (minimal planned/preventive maintenance). Breakdowns occur because of sudden/sporadic failures of different components/equipment (aim for zero breakdowns) • Lack of control on maintenance cost, spare parts and equipment losses • Measurement of deterioration and restoration of components (needs condition-based maintenance practices) • No spare part maintenance management system (need application of spare parts maintenance management system in practice)
Quality Management and Improvement	• Minimum focus on Quality Circles, Zero defects philosophy • Quality tools such as & 7 New and & 7 old tools seldom practiced • Minimal manufacturing process data being used (although very important) • Nominal 3 M (man, material and method) focus on quality issues (machine is outside the focus) • No plan for improvement (need team approach to solve improvement projects using quality tools), and • TPM tools such as why–why, and performance measurement analyses were not in use for individual learning (need use of these to address problems)
Education and training	• For operators, only production process training (not enough technical skill or AM training) • For technicians, basic technical/equipment training (not for problem-solving) • No training on various quality tools, i.e., SPC, Six Sigma, 7 New and old Quality tools • For engineers/managers, equipment and (no training on TPM/quality (sigma)) management tools • Least emphasis on education and problem-solving strategy using cross-functional, i.e., AM, PM and FI teams

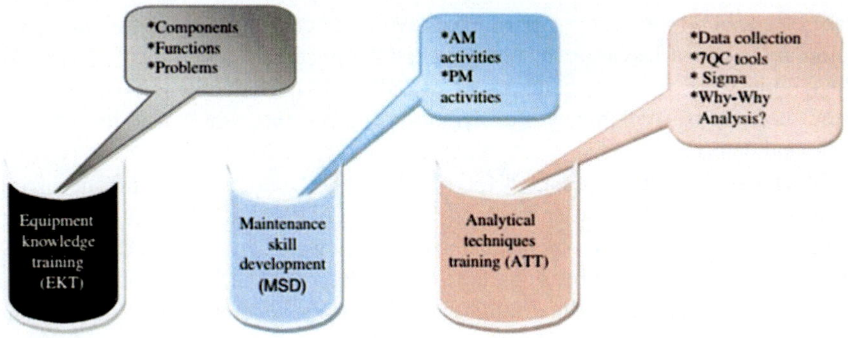

Fig. A.2. Training modules

Training Program

Training program (Fig. A.2) consisting of three main modules, i.e., (i) Equipment knowledge training (EKT), (ii) MSD (Maintenance skill development) and (iii) Analytical techniques training (ATT) were designed to upgrade the skills of operators/technicians in AM, FM, FI teams and members from BB, GB, YB teams. Each module consists of various tools and techniques.

Measure Phase

To account for maintenance-related aspects in the cell, data are collected for a period of about 6–8 months from the historical records in maintenance logs. The total failure causes reported were 1300. From the statistics presented in Table A.2, it is observed that in case of human failures, the MTTR is 1.28 h with MTBF 20 h. Failures in this category included any stoppage of the system attributed to improper actions (such as pushing the wrong button or lever, using the wrong weight of oil, failure to take the correct remedial actions such as not closing an interlocking door or not tightening a bolt). Also, maintenance-related human failures may be due to errors of omission or errors of commission, i.e., lack of attention, confusion in

Table A.2 Failure and repair statistics

S. no	Failure causes	MTBF (h)	MTTR (h)
1	Human failures	20.00	1.28
2	Mechanical failures	54.20	1.18
3	Hydraulic failures	38.17	2.00
4	Electrical failures	30.20	0.40
5	Electronic failures	92.80	0.75

cables (put in wrong order), use of excessive force causing instrumentation cables to break or use of less force resulting in bad connections, loose bolts, etc. 15% of the failures were hydraulic failures with MTBF 38.17 h. Hydraulic failures normally occur because of contamination of hydraulic components, which lead to increased wear and deterioration.

Mechanical failures accounted for 17.50% of the total, with failures occurring at an average rate of one after every 54 h and the MTTR such failures was 1.18 h. Untimely stoppages caused by mechanical failures were associated with wearing of components such as gears, bearings and tooling. 10% of all the failures were classified as electrical failures. Electrical failures displayed the second shortest MTBF, one failure after every 30.20 h. The electrical failures required, on average, the least amount of time to repair, i.e., 0.40 h. Blown fuses or dirty limit switches and failures of the electromechanical components such as motors, relays, starters and wiring were the primarily reasons for electrical failures. Electronic failures that normally consist of failures of solid-state components such as logic buses, power supplies and servo drives accounted for only 6% of all failures with the highest MTBF, i.e. 92.80 h requiring an average repair time of 0.75 h. As mentioned earlier that on the quality front, Paper machine I and Paper machine II were working on 2.56 and 2.84 sigma levels, respectively, which were quite low.

Based upon the monthly production and rejection report (for machine—II) assuming 1100 tons of monthly production a total of 273.02 tons is rejected due to various types of defects such as wrinkle, high cob, dye spots, dull shade, others (GSM, i.e., gram per square meter, bursting strength, tear factor, ash content, moisture content, etc.). The schematic representation of these defects is shown in Fig. A.3.

The collected data show that the rejection in the process was 248,200 PPM. The corresponding *sigma rating* of the process can be approximated to 2.18.

The schematic representation of these defects is shown using Pareto chart below. DPMO level of process was found to 248,200 PPM with the corresponding sigma rating approximated to 2.56. To determine overall distribution of % defective for the

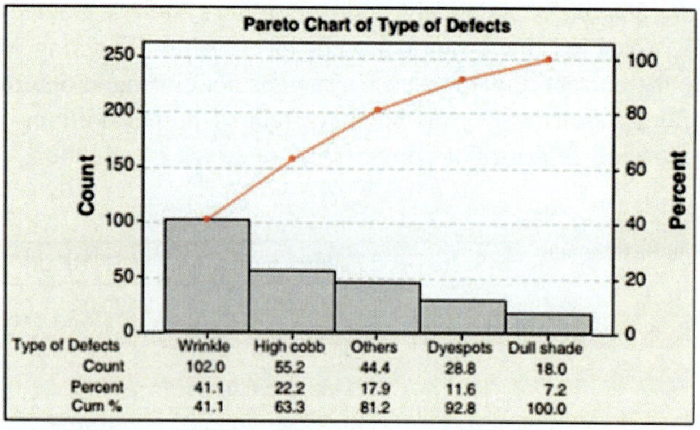

Fig. A.3. Pareto chart

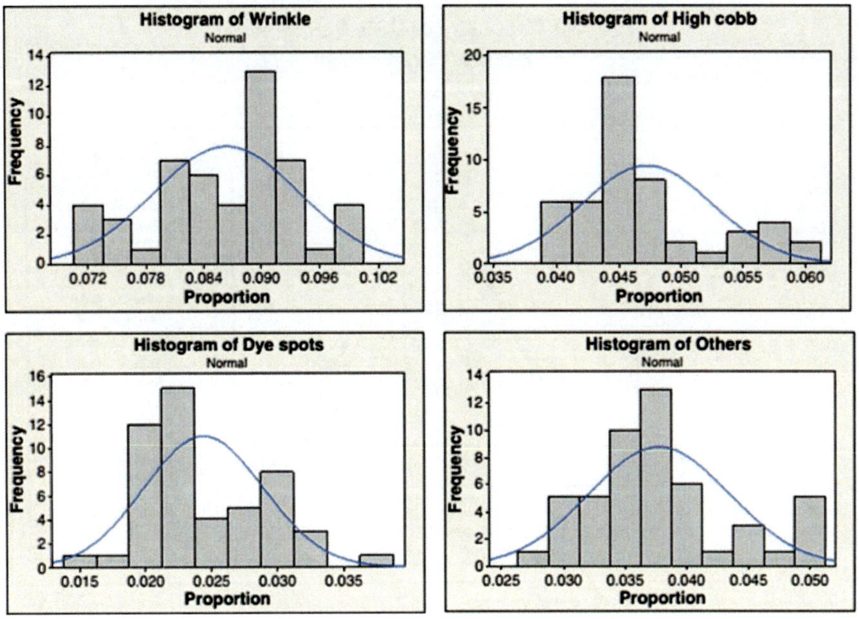

Fig. A.4. Histogram plots for defects

defects, histograms for wrinkle, high Cobb, dyespots and others are drawn as shown in Fig. A.4.

Based upon the data collected under other defects, data of GSM variation account for major rejections by the customers. It was decided to conduct process capability study (Fig. A.5) to compare the process output with customer requirements. The Minitab statistical software output of process capability evaluation is presented in Fig. A.5. As $C_{pk} = 0.40 < 1$, process cannot reach specification limits. Process average value is away from the target value. Also, Z-Bench sigma value of process was found out to be 1.06, and existing DPMO level of the process comes out to be 144,958.98, which is remarkably high, and this shows that there are a lot of opportunities for improvement in the process.

Analysis Phase

Brainstorming session was conducted by the team leaders to diagnose the unreliable aspects of the paper machine and all the possible failure causes related to the machine units, i.e., forming, press and dryer are listed out using RCA diagram as shown in Fig. A.6. Further, team leaders decided to conduct FMEA in order to identify potential system failure modes, their causes and effect on performance of the system. Failure mode and effect analysis of all the units, i.e., forming, press

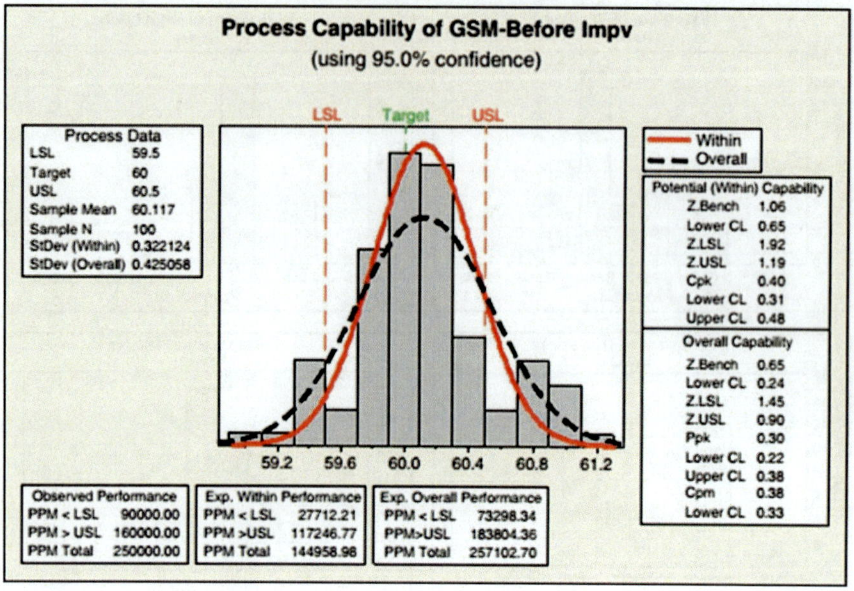

Fig. A.5. Process capability plot

and dryer, is conducted by breaking the unit into its subunits. For instance, Table A.3 presents FMEA of forming unit. The methodology used to compute the scores related to failure of occurrence (O_f), likelihood of non-detection of failure (O_d) and severity (S) of failure of various components is discussed in detail.

Based upon FMEA analysis (Tables A.3 and A.4), the different failure modes associated with forming, press and dryer components were identified and are presented in pie chart as shown in Fig. A.7. Vibrations/noisy operation (steam rollers)—13%, foreign materials (fiber mat, grit, nails)—12%, bearing seizure (vacuum pump)—12%, jamming, breakage of baffles—9% were identified as one of the main failure modes. The recognition of potential failure modes associated with each type of failures (mechanical failures, hydraulic failures, electronic failures, electrical failures, human failures and software failures) helped the members from AM, PM and FI teams to initiate investigations of recurring equipment/maintenance failures, repair/replacement costs and associated corrective actions.

Improvement and Control Phase

The aim of this phase is to examine the reasons that appear during the analysis phase and devise improvement/control plans in order to reduce them. Table A.4 presents the improvement/control plans based on FMEA analysis. The team leaders and group leaders discussed various improvement plans suggested by the operators with TPM

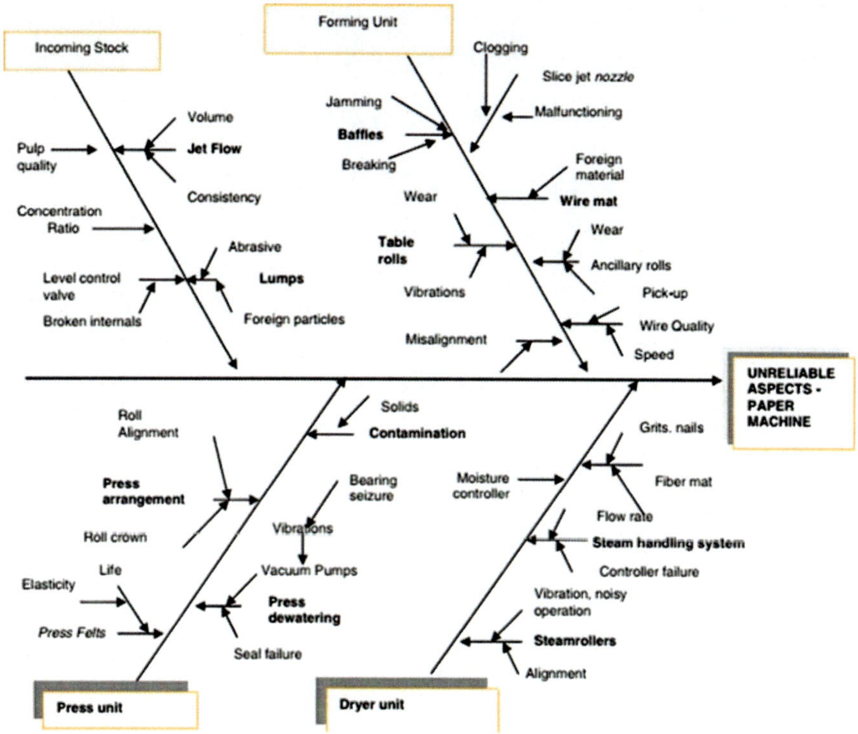

Fig. A.6. Cause and effect diagram

and Six Sigma managers. Then, with the help of personnel from production and design department, feasibilities of such plans were worked out. Depending upon the feasibility, the plans were incorporated. In order to facilitate AM personnel for daily maintenance routine interventions and recognize various trouble shootings (with the help of FI teams) associated with failures of critical components, the computerized defect recording system was developed. It consists of complete record of commonly failed/repaired or replaced components associated with corresponding machining centers with details such as failure detection date time, mean repair time and MTBF. Initially, the maintenance program based on Nakajimas seven steps of AM was followed.

- Initial cleaning
- Countermeasures for cause and effects of contamination sources
- Cleaning and lubrication standards
- General inspection
- Autonomous inspection
- Organization and tidiness
- Full implementation of autonomous maintenance

Table A.3 FMEA for forming unit

Component	Function	Potential failure mode	Potential effect of failure	Potential cause of failure	O_f	S	O_d	RPN
Head box – Baffles – Perforated plates/bars – Slice jet/nozzle – Level control valve	*To discharge pulp on to wire.* To regulate the level	Breaking Jamming Blockage Fails to open Fails to open fully Fails to close	Non-uniform interrupted flow Unable to provide required conc. (pulp+ water) Failure to provide full-metered flow Loss of flow	Broken internals [FC_{11}] Corrosion [FC_{12}] Scale buildup [FC_{13}] Particulate contamination [FC_{14}] Mechanical binding [FC_{15}] Scale building [FC_{16}] Broken internals [FC_{17}]	3 4 6 7 4 5 3	7 5 5 7 7 8 6	4 5 8 8 6 7 7	84 100 240 392 168 280 126
Fourdinier wire Table Wire mat – Table rolls – Suction rolls – Dandy rolls – Cough rolls – Pickup rolls	Carry the pulp Support the wire Drainage of water Run freely on the surface Dewatering the pulp Transfer sheet to pickup felt	Abrasion Building of fiber mat Misalignment Buckling/deformation Looseness Sagging Breaks Bearing seizure Breaks Bearing seizure	Holes /marks on the sheet Rapid wear and shorten the life Variation in wire tension Loss in operation Stock jumps and creates disturbance on wire Sheet formation interrupted (crush and curl) Fails to transfer sheet to pickup felt	Foreign materials [FC_{21}] (sand, grit, nails etc.) Lumps/pimples etc. [FC_{22}] Roll wear [FC_{23}] Vibrations [FC_{24}] Out of balance [FC_{25}] Improper maintenance [FC_{26}] Mechanical stresses [FC_{27}] Bearing seizure [FC_{28}] Jammed shafts [FC_{29}] High temperature [FC_{210}] Nip pressure [FC_{211}] Vibrations [FC_{212}] High temperature [FC_{213}]	8 7 4 5 6 3 7 6 6 5 6 4 4	10 9 9 7 6 7 8 7 8 8 8 9 8	9 8 8 8 5 5 5 6 8 8 4 8 9	720 504 288 280 180 105 280 252 384 320 192 288 288

(continued)

Table A.3 (continued)

Component	Function	Potential failure mode	Potential effect of failure	Potential cause of failure	O_f	S	O_d	RPN
Suction Box Vacuum pumps	Complete dewatering/ drainage	Fails to operate continuously	Piston fails to execute the movement	Breaking of piston rod [FC_{31}]	4	9	9	324
		Leakage from casing	Air may enter the system	Seal failure [FC_{32}]	7	6	8	336
		Rotor jamming	Pump motor overloaded	Excessive radial thrust [FC_{33}]	6	7	4	168
			Pressure loss	Lack of lubrication in moving parts [FC_{34}]	4	8	6	192
				Bearing failure [FC_{35}]	6	7	6	252
				Inclusion of solid particles into clearances [FC_{36}]	6	6	7	252

Table A.4 FMEA and improvement/control solutions

Component	Potential failure mode	Potential effect of failure	Potential cause of failure	Improvement/Control actions
Head box – Baffles – Perforated plates/bars – Slice jet/nozzle – Level control valve	*Breaking* *Jamming* **Blockage** Fails to open Fails to open fully Fails to close	*Non-uniform interrupted flow* Unable to provide required conc. (pulp+ water) Failure to provide full-metered flow Loss of flow	Broken internals Corrosion Scale buildup Particulate contamination Mechanical binding Scale building Broken internals	Replace broken internals Corrosion monitoring Remove scale build-up Check contamination Check and imitate action Remove scale build-up Replace broken internals
Fourdinier wire Table – Wire mat – Table rolls – Suction rolls – Dandy rolls – Cough rolls – Pickup rolls	Abrasion Building of fiber mat Misalignment Buckling/deformation Looseness Sagging Breaks Bearing seizure Breaks Bearing seizure	Holes /marks on the sheet Rapid wear and shorten the life Variation in wire tension Loss in operation Stock jumps and creates disturbance on wire Sheet formation interrupted (crush and curl) Fails to transfer sheet to pickup felt	Foreign materials (sand, grit, nails etc.) Lumps/pimples etc.] Roll wear Vibrations Out of balance Improper maintenance Mechanical stresses Bearing seizure Jammed shafts High temperature Nip pressure Vibrations High temperature	Initiate 5S and AM practices Check wear and initiate condition monitoring Install PM actions Initiate AM and PM actions Initiate CBM Check pressure and temperature
Suction Box Vacuum pumps	Fails to operate continuously Leakage from casing Rotor jamming	Piston fails to execute the movement Air may enter the system Pump motor overloaded Pressure loss	Breaking of piston rod Seal failure Excessive radial thrust Lack of lubrication in moving parts. Bearing failure Inclusion of solid particles into clearances	Replace Check and Replace Check leakage Perform AM actions Cleaning lubrication

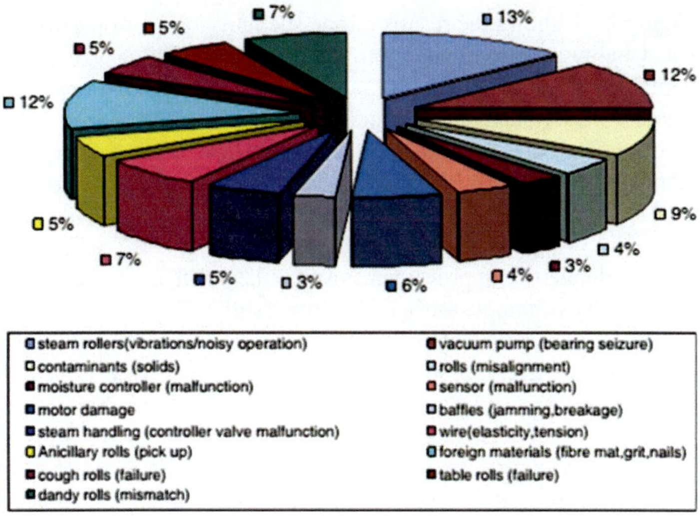

Fig. A.7. Failure modes

On the part of Six Sigma implementation, the following improvement actions were initiated by Six Sigma teams (i) CI projects: Though the full implementation of framework takes a few years after inception, if the improvement process is started in a systematic manner, then the results of successful implementation can be visualized within 3–4 years.

The number of CI projects (also known as Kaizens) year wise with target and actual values and finally their execution are shown in Fig. A.8a. It is observed that to be successful, the team approach is more effective, and participation by the members in brainstorming sessions for problem-solving helps to achieve the targets. (ii) Employee skill improvement projects. To help the operators and maintenance personnel to learn more about equipment functions, common problems, their occurrence and the ways to prevent these problems, a large number of skill development and skill transfer projects in different areas had been undertaken and executed in

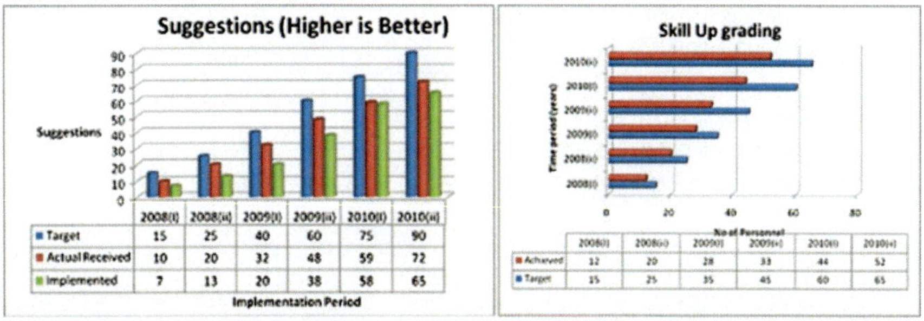

Fig. A.8. Kaizens and skill upgrading

the cell. Figure A.8b shows the results of one such program for operators and technicians. (iii) Quality improvement. By following the quality management practices and following defect identification program (check sheets, histograms, Pareto analysis), and problem analysis (control charts and run charts, cause and effect diagrams) with the help of members from green, black and yellow belt teams, a considerable improvement in GSM variation has been observed as shown in process capability plot in figure with $C_{pk} = 1.81 > 1$ and Z-Bench sigma value 5.43 with DPMO level 0.03.

In order to validate the improvement results, the verification analysis has been carried out through two sample t-test as shown in Table A.5. The t-test supports the significance of the difference in the means of before and after implementation of Six Sigma framework. The sig. (two-tailed) P value obtained is = 0.000 for all three. As this value is <0.05, it is concluded that there is statistically significant difference between the means exist. The descriptive statistics is presented in Table A.5, respectively.

The comparative analysis between before and after the Six Sigma implementation for the defects, i.e., wrinkle, high cob and others prioritized using Pareto analysis in measure phase is graphically shown using the box plots in Fig. A.9a–c.

Table A.5 Descriptive statistics for defects

Defects type	N	Mean	St Dev	SE Mean
Wrinkle (Before Six Sigma)	50	0.08637	0.00753	0.0011
Wrinkle (After Six Sigma)	50	0.02138	0.00138	0.00020
Difference = mu (Wrinkle (Before Six Sigma)) − mu (Wrinkle (After Six Sigma)) Estimate for difference: 0.06499 95% CI for difference: (0.06081, 0.06516) T-Test of difference = 0 (vs. not =): T-Value = 58.19 P-Value = 0.000 DF = 52				
High Cobb (Before Six Sigma)	50	0.04728	0.00528	0.00075
High Cobb-(After Six Sigma)	50	0.01298	0.00284	0.00040
Difference = mu (High Cobb (Before Six Sigma)) − mu (High Cobb-(After Six Sigma)) Estimate for difference: 0.034301 95% CI for difference: (0.031612, 0.034990) T-Test of difference = 0 (vs. not =): T-Value = 39.28 P-Value = 0.000 DF = 75				
Others (Before Six Sigma)	50	0.03777	0.00569	0.00081
Others (After Six Sigma)	50	0.01748	0.00234	0.00033
Difference = mu (Others (Before Six Sigma)) − mu (Others-(After Six Sigma)) Estimate for difference: 0.018290 95% CI for difference: (0.017551, 0.021029) T-Test of difference = 0 (vs. not =): T-Value = 22.15 P-Value = 0.000 DF = 65				

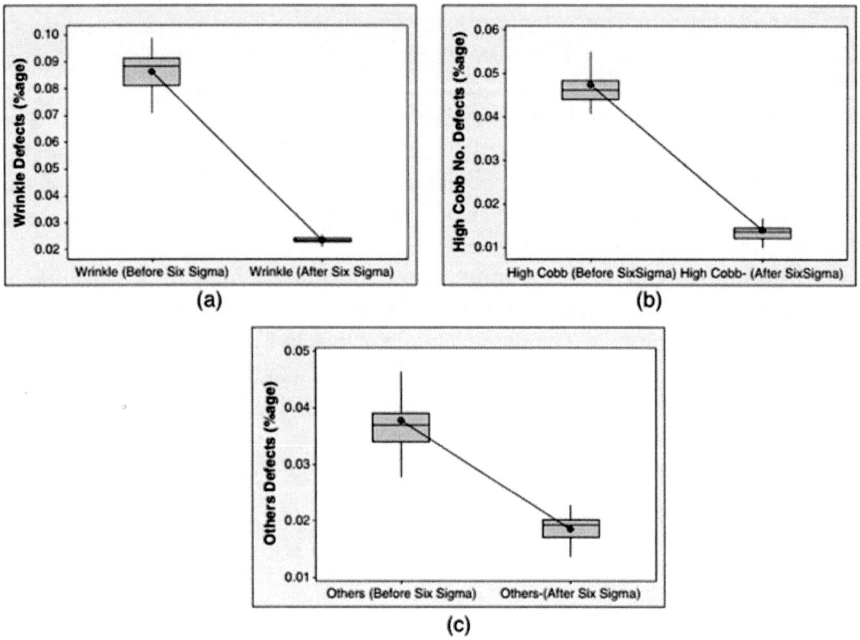

Fig. A.9. **a–c** Comparative analysis between before and after the Six Sigma implementation

Intangible Benefits

(i) **Introduction of concept of total quality**: With aim for zero defects through management of 4 M conditions all members/operators were responsible for maintaining equipments/facilities by performing AM and PM activities (cleaning, inspection and lubrication). The approach had helped in spreading and growing the concept of total quality through TPM and Six Sigma interventions among all in the cell which not only helps in upkeep of equipment but also improves the quality level and reduces production losses in terms of scrap and rework.

(ii) **Setting up of cross-functional teams**: Implementation of TPM + Six sigma culture in the cell had helped to form cross-functional teams consisting of team members from maintenance and production departments [(AM), (PM) and (FI) + sigma belts (GB, BB, YB)]. Thus, allowing members to identify and resolve many basic equipment and quality problems with in short period of time.

(iii) **Training and skill development**: The training modules (EKT, MSD and ATT) that were specifically designed to help operators to learn more about how their equipment functions? What common problems can occur?, Why they occur?, and how these problems could be prevented? had helped a lot in training and upgrading the skills. Handling of simple repair works with the assistance of FI personnel had nurtured necessary maintenance skills in the operators for solving the problems without causing any further delays, thus reducing mean time to repair, i.e., MTTR. Knowledge of QC tools helps the members from sigma belts to collect and analyze the problems related with quality.

(iv) **Increased responsibility**: Both cross-functional teams consisting of team members from maintenance and production departments [(AM), (PM) and (FI) accepted the responsibility for equipment condition. Now they not only concentrate on production but also perform simple maintenance tasks. A change in the traditional syndrome "I make—you fix," was observed.

(v) **Development of sense of importance for maintaining basic equipment conditions**: The sense of increased responsibility after implementation of TPM program had promoted more operator involvement by performing AM (oiling, lubrication, inspection) and 5S housekeeping activities. This had helped to generate the sense of importance for maintaining basic equipment among the operators (Table A.6).

Table A.6 Summarized results

	Before TPM+Six Sigma	After TPM+Six Sigma	Remarks
Machine failure rate (PM-I)	7.49	2.4	Lower is better
Machine failure rate (PM-II)	6.73	2	Lower is better
Rework	22%	10%	Lower is better
AOQL	30	5	Lower is better
Z (Bench sigma) GSM	1.06	5.43	Higher is better
Sigma level (PM-I)	2.56	4.01	Higher is better
Sigma level (PM-II)	2.84	4.15	Higher is better
Maintenance versus Operation cost	30%	10%	Lower is better
Accident rate	1.5	0.5	Lower is better
Defect rate	20%	5.40%	Lower is better
Customer Complaints	24%	8%	Lower is better
Availability (PM-I)	0.82	0.91	Higher is better
Availability (PM-II)	0.84	0.93	Higher is better
Performance efficiency (PM-I)	0.78	0.905	Higher is better
performance efficiency (PM-II)	0.81	0.92	Higher is better
Quality rate (PM-I)	0.75	0.95	Higher is better
Quality rate (PM-II)	0.8	0.97	Higher is better
OEE (PM-I)	49	78	Higher is better
OEE (PM-II)	54	80	Higher is better

Conclusions

It is observed from results that considerable improvements in availability, OEE and quality level (AOQL) are observed after implementation of TPM program with help of cross-functional teams. A significant improvement was observed in the key performance metrics MTBF, OEE, AOQL level and defect rate, etc. as shown in figures. The implementation of framework provided an impetus for establishing best practices within the company. Standard housekeeping procedures helped to reduce the number of accidents at workplace significantly. Implementation of such framework not only helps increases in efficiency and effectiveness of manufacturing systems but also prepares the company to meet the challenges put forward by globally competing economies to achieve world-class manufacturing status.

The study attempts to integrate two well-established strategies aimed at improving the plant performance so that it will yield a great value to practicing engineers, academicians and consultants working in the field of quality and maintenance management. The outcome of the present study opens up a new perspective for attaining competitive edge among business houses through cultural change. Both TPM and Six Sigma are key business process strategies, which can be employed by companies to enhance their manufacturing performance. In a nutshell, we can conclude that the benefits of the said operational integrated framework are achieved by

- Forming cross-functional teams consisting of operators, maintainers, engineers and managers to improve individual employee and equipment performances
- Increasing operator involvement and ownership of the process through autonomous maintenance paradigm
- Improving problem-solving skills of operators by demonstrating the improvements using statistical tools such as Pareto Chart, Box plots, FMEA and control charts
- Refining preventive and predictive maintenance activities
- Focusing on reliability and maintainability aspects of equipment
- Focusing on quality problems through sigma teams (BB, GB, YB)
- Facilitating and practicing problem-solving through team spirit
- Promoting leadership and management support for improvement programs
- Improving morale of teams through reward and incentives
- Imparting training and coaching to BB, GB, YB teams on statistical methods
- Adopting improved procedures (in form of kaizens) for:

    (i) changeovers and setups,
    (ii) carrying out maintenance tasks, and
    (iii) better training of operators and maintainers.

Appendix B
Case Study 2

#Attaining Competitive Positioning Through SPC—An Experimental Investigation from SME

Summary

The purpose of this case study is to provide a conceptual framework that connects theory with straightforward application of statistical process control (SPC) in discovering and analyzing causes of variation to eliminate quality problems, which not only helps small and medium enterprises (SMEs) to improve their processes but also helps to attain competitive positioning.[2] Based on theory and methodological framework, an experimental study has been presented. Use of histograms, X (bar) and R control charts and process capability plots and cause-and-effect diagrams have been made to analyze the assignable causes. A case from an SME engaged in machining of automotive parts is investigated. The results demonstrate the effectiveness of SPC in evaluating and eliminating quality problems. The machine capability (CP) and the process capability (CP_k) values are also obtained to know inherent variation in the process. If these quality tools are applied with management support and apt knowledge, attained through proper training and motivation, then in this cut-throat competitive world, SMEs can establish their market position by enhancing the quality and productivity of their products/processes. From the study, the authors conclude that application of SPC requires thorough preparation, management commitment and human resource management through proper training, teamwork and motivation embedded with a sound measurement and control system. An attempt has been made

[2] https://www.emerald.com/insight/content/doi/10.1108/MBE-10-2013-0050/full/html.

R .K. Sharma, *Quality Management Practices in MSME Sectors*,
Springer Tracts in Mechanical Engineering,
https://doi.org/10.1007/978-981-15-9512-7

to bridge the gap between theory and practice by developing a conceptual framework and providing a practical support by illustrating a case from an SME engaged in machining of automotive parts.

 # Sharma Rajiv, Kharub Manjeet (2014) "Attaining Competitive Positioning Through SPC—An Experimental Investigation from SME", Measuring Business Excellence, Vol. 18 Issue 4 pp. 86–103.

1. Conceptual Framework for Statistical Process Control (SPC) Implementation

Though small and medium enterprises (SMEs) operate in different business contexts from large business houses, but some elements of management practices are applicable to all organizations irrespective of their size. Statistical process control (SPC) has been found to be a valuable technology for understanding process dynamics and is used as an integral part of monitoring, managing, maintaining and improving the performance of a process through the effective use of statistical methods (Elg et al. 2008). The conceptual framework for *straightforward application of statistical process control (SPC) is* shown in Fig. B.1a and b. It consists of five phases, i.e., (i) preparation, (ii) definition, (iii) measurement, and (iv) control phase.

- *First phase* deals with introduction to prerequisties, i.e., top management commitment, engineering aspects and meeting human resource management needs, which are supposed to be essential for launching of SPC program in any organization. Formation of organizational structure consists of (i) top management representative, (ii) a steering committee, and (iii) process action teams, i.e., PATs for initiating small pilot projects is the next step in this phase.
- *Second phase*, i.e., Definition phase helps the PATs to clearly define the problem definitions related to product/product assurance, process/process control. These definitions form the basis for specifications and to kick-off control activities critical to a process during production.
- *Third phase* is the measurement phase in which the process is described and the boundaries are determined. The measurement system should focus on integrated process control, i.e., off-line and on-line process control. Use of control charts is done to control incoming materials, machine characteristics, human factors to detect process shift because of assignable causes. Follow product assurance by verifying conformance to requirements by sorting out/scrapping defectives. Process capability studies can be used to measure and analyze the process performance. Another important step in this phase is selection and design of control charts, more details can be had from Oakland (1999).
- *Fourth Phase* is the control phase in which process monitoring and control chart interpretation are carried out. Table B.1 presents generic rules that are proved useful for interpretation (Montogomery 2005).

Out of control situations can be investigated by discovering possible assignable causes. The PATs should come up with root cause analysis or brainstorming sessions to note down causes of inherent variation. The potential solutions for out of control situation can be found and prioritized for initiating corrective actions. Further, any

Table B.1 Rules for interpretation of control charts

A process is said to be out of control	Description of rules
	i. If a point falls outside the control limits
	ii. If two out of three successive points fall outside the warning limits on the same side of the center line. Warning limits are placed at two standard deviations from the center line
	iii. If four out of five successive points fall outside one sigma limit on the same side of the center line
	iv. If seven or more successive points fall on one side of the center line
	v. If there is a run of seven or more successive points either above or below the center line
	vi. If the chart shows periodic low and high points also called a cyclic pattern of variation

more out of control situations are analyzed and if not found process maintenance and documentation can be done. Once the process stability is achieved, the next step is to go for process capability analysis. Process capability is measured by the process capability index, *Cp*, which is computed as the ratio of the specification width to the width of the process variability.

$$C_p = \text{Specification width (USL} - \text{LSL)/Process width } (6\sigma)$$

Where; the specification width is the difference between the upper specification limit (USL) and the lower specification limit (LSL) of the process. The process width is computed as 6 standard deviations (6σ) of the process being monitored. As in case of (6σ) spread, most of the process measurement (99.74%) falls within $\pm$ 3 standard deviations, which is a total of 6 standard deviations.

- $C_p = 1$: *A value of* C_p *equal to* 1 *means that the process variability just meets specifications, we would then say that the process is minimally capable.*
- $C_p \leq 1$: *A value of* C_p *below* 1 *means that the process variability is outside the range of specification. This means that the process is not capable of producing within specification and the process must be improved.*
- $C_p \geq 1$: *A value of* C_p *above* 1 *means that the process variability is tighter than specifications and the process exceeds minimal capability*

The process capability index C_{pk} additionally considers the position of the distribution. Possible lack of centering of the process over the specification range is measured using C_{pk}

$$C_{pk} = \min(\text{USL} - \mu/3\sigma),\ (\mu - \text{LSL})/3\sigma$$

Table B.2 Process estimation based on as: C_p and C_{pk}

Capability index	Estimation of process
$C_{pk} = C_p$	Process placed exactly on the center of field of tolerance
$C_p < 1 \rightarrow T = 6\sigma$	Process is not stable: Necessary is "improving" of process or widening of tolerance
$C_p = 1 \rightarrow T = 6\sigma$	Process is minimally capable
$C_p > 1.33 \rightarrow T > 8\sigma$	Capability of process is good: If $C_p = C_{pk}$
$C_p > 1.66 \rightarrow T > 10\sigma$	Capability of process is very good
$C_{pk} \neq C_p$	Process unable to work, wrong putting: Required is new putting of process

where: μ the mean of the process σ the standard deviation of the process. It evaluates the critical distance between process level and the upper and lower specification limit.

When the process is centered between these limits, $C_{pk} = cp$. A deviation from the centered position causes $C_{pk} < c_p$.The industry usually asks for c_p and $c_{pk} > 1.33$, which is specified in contracts between customer and the supplier. Table B.2 provides the examples of process estimation based on both c_p and C_{pk} index. The potential benefits of SPC program should be evaluated in terms of improvement in quality, i.e., reduction in scarp/rework/rejects. This helps in wider acceptance and companywide implementation of SPC becomes easy with the support of all from top to bottom.

2. **Illustrative Case**

Company considered in the study (SME situated in northern part of India) was facing increase in rejections with respect to one of the automotive part, i.e., crankcase being supplied by its business partner. The part is being machined to proper dimensions before finally used in assembly work. After having examined the rejections, it was found that there were some problems with sizes of 62 + 0.019, 30 + 0.038, 42–0.017/−0.037, 21.5 + 0.033, 32–0.017/−0.042 and 35–0.017/−0.042, respectively. The company was using a traditional approach, i.e., inspection-based quality control to inspect the final product and screen out items not meeting the specifications. The sizes that do not meet the desired specifications are sent back to the machines for remachining, otherwise are scrapped. This results in increase in scrap rate, customer complaints, and decreased productivity. Though management knew about SPC, but earlier initiatives on SPC fail to perform adequately that might be because of lack of recognition of the technique and its relevance within the organization. Therefore, before implementation of the proposed framework, it was decided to review the present problems that lead to failure of SPC initiatives in the company. Table B.3 presents the summary of reasons responsible for failure of SPC initiatives.

Table B.3 Summary of reasons responsible for failure of SPC initiatives

Management commitment	• Constant attention and support of top management were not observed in quality improvement programs • Management principles such as delegation of tasks, responsibilities and authority to the lowest possible level were not followed
Quality awareness and education and training	• Least emphasis on education and problem-solving strategy using cross-functional teams • Lack of statistical orientation • Engineers/managers are not well versed with data collection, analysis and design procedures of control charts • No training on various quality tools, i.e. SPC, Six Sigma,7 New and old Quality tools • For operators, technicians no training on defect prevention
Quality improvement initiatives	• Minimum focus on Quality Control Circles, Zero defects philosophy • No plan for improvement (need team approach to solve improvement projects using quality tools) • Minimal manufacturing process data being used (although very important) • Quality tools such as & 7 New and & 7 old tools seldom practiced • Nominal 3 M (man, material and method) focus on quality issues (machine is outside the focus) • Lesser communication with customers, and suppliers with respect to product, specifications and delivery performance
Material control	• More stress on final inspection, though defects because of materials can be controlled by incoming inspection, and by in process inspection
Machines and tools	• Mostly corrective maintenance is performed. Though proactive maintenance strategies (condition monitoring) along with preventive maintenance (periodic repair/overhaul) on machines may be performed
Measurement tools and process setup	• A preventive approach to calibrate and performing gage repeatability and reproducibility (R&R) study was not adopted • Seldom use of control charts to investigate the stability of the measurement tools is done • Need to build and practice procedures for process setup
Teamwork	• Various departments of company lacks teamwork
Pilot projects	• Least emphasis on small projects

2.1. SPC Implementation Steps

After summarizing the problems, management felt the need to build an organization structure to implement SPC in the organization. The structure consists of (i) top management representative, (ii) a steering committee, and (iii) process action teams, i.e., PATs. Production manager was made coordinator of the steering committee and managers from different departments, i.e., procurement, design and development, quality and maintenance were appointed members of the steering committee. A process action team includes operators, supervisor(s), process engineer(s), quality engineer(s), and an SPC expert/facilitator. Steering committee was entrusted with tasks of (i) stimulating SPC awareness through total employee involvement (iii) team-building and recognition (iv) providing method and means for process/performance measurement (iv) initiate relevant training support (iv) provide budget provisions

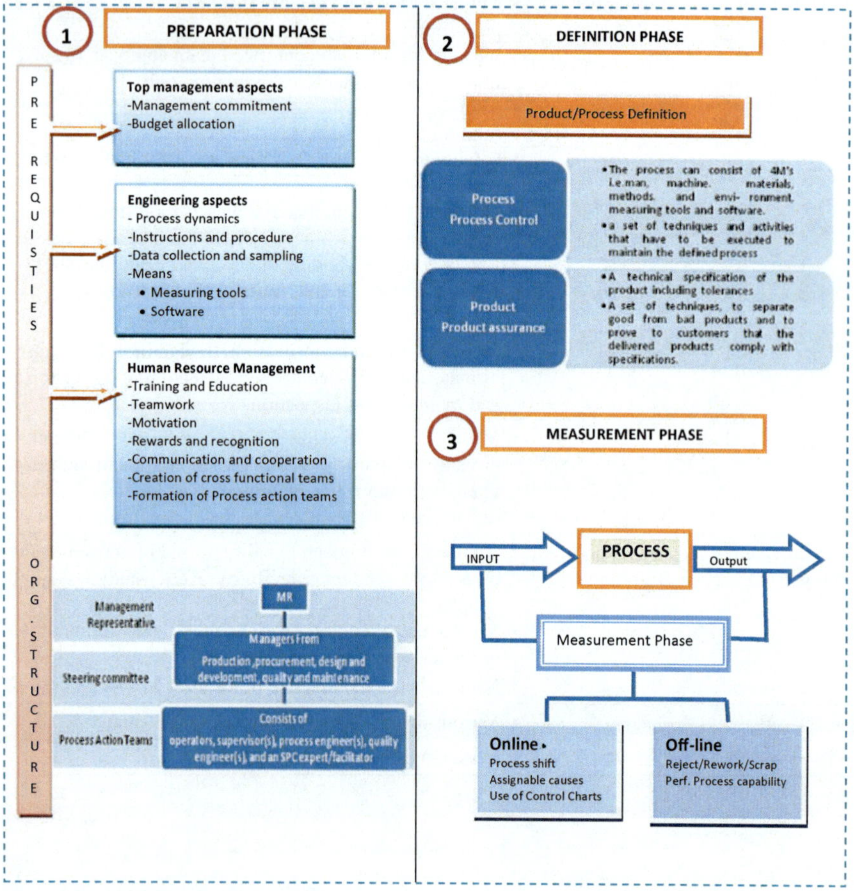

Fig. B.1 **a** Preparation, definition and measurement phases **b** measurement (contd-) and control phase

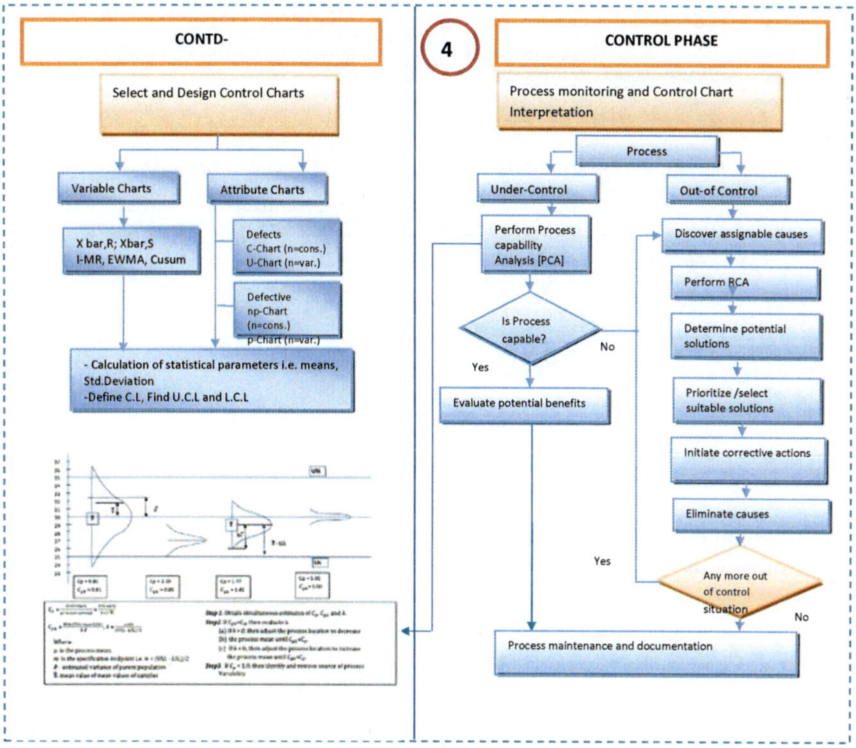

Fig. B.1 (continued)

to initiate pilot projects and realize improvements (iv) monitor progress and assess results of PATs (v) set priority for quality management activities and make sure that the control plan is developed (vi) reporting to top management about the project progress and outcomes. The main task of the PATs is to bring the process under control by following four steps 1. Process/Problem definition 2. Perform root cause analysis 3. Defining measurements framework, i.e., use of suitable control charts 4. Performing Process capability studies.

- ***Step 1: Process description***: In this step, the process is described and the boundaries of the process under study are determined.
- ***Step 2: Description of activities***: For the process described in step 1, the main problems (causes) and their effects are listed using cause and effect diagram commonly known as Ishikawa or Fish-bone diagram. All possible causes that could influence quality characteristics and ultimately the machining accuracy is represented by fishbone diagram as shown in Fig. B.2.
- ***Step 3: Define measurements framework***: In this step, a suitable measurement method to discover and eliminate the root causes for problems is adopted. The team members are entrusted with selection of the parameters for controlling the

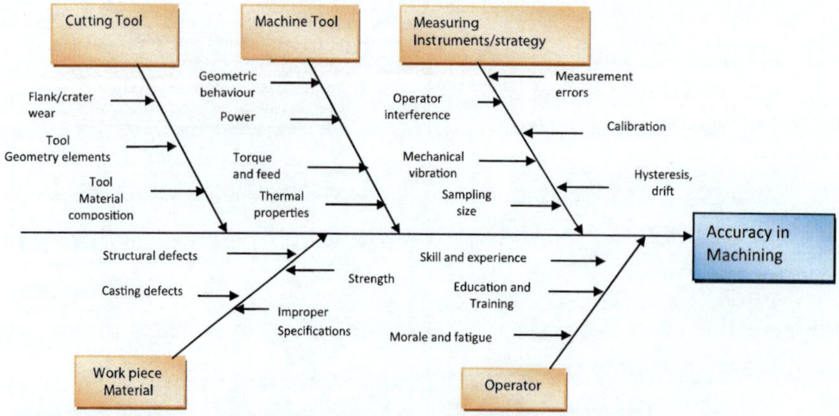

Fig. B.2 Cause and effect diagram for machining inaccuracy of crankcases

process output. Further, a plan is made to collect, to monitor, and to analyze the measurements.

- ***Step 4: Performing process capability studies***: To measure the level of statistical and technical control of the process or in other words to judge whether the level of control is satisfactory to meet specification limits, process capability studies are performed. Process capability indices are calculated to quantify the ratio between tolerance width and inherent process variation (C_p) and the effect on this ratio due to a deviation of the process mean from the target value (C_{pk}) index.

3. **Results and Discussions**

The machining data of crankcase dimensions shown in Table B.4 are categorized as shift data set-I. Similarly, data related to other two shifts were categorized as shift data set-II and shift data set–III, respectively. First the histogram plots were obtained to check the frequency distribution. Further, control charts (X bar and R) were drawn for each part and process capability values, i.e., C_p and C_{pk} are obtained to know inherent variation in the process. Three different shift data sets pertaining to crankcase dimensions were analyzed to see the application of statistical quality control. Table B.4 presents the diameter dimensions of Data set-I.

For each shift, the statistical parameters (average, range, etc.) were calculated to enable further process capability analysis and construction of the control charts. First, the histogram plot for dimension (62 + 0.019) is drawn and, X bar- R control chart is drawn as shown in Fig. B.3. Further in order to know the inherent capability of the process, process capability plot is obtained as shown in the Fig. B.4. Similarly, for all the dimensions of the data set 1, the observations based on X bar-R control chart and process capability plots were drawn. The average, maximum size, minimum size, ranges, standard deviation, six sigma values and process capability value so

Table B.4 Dimensional Data Set-I

S. no	Dia 62 + 0.019		Dia 42–0.017/-0.037		Dia 30 + 0.038	
	Min	Max	Min	Max	Min	Max
1	62.011	62.013	41.976	41.978	30.020	30.024
2	62.010	62.014	41.975	41.977	30.019	30.023
3	62.009	62.012	41.976	41.980	30.020	30.023
4	62.012	62.013	41.971	41.977	30.021	30.023
5	62.010	62.012	41.976	41.980	30.021	30.022
6	62.010	62.013	41.977	41.981	30.020	30.024
7	62.010	62.012	41.970	41.976	30.021	30.023
8	62.011	62.013	41.972	41.977	30.019	30.024
9	62.010	62.012	41.971	41.976	30.018	30.023
10	62.009	62.011	41.972	41.975	30.018	30.021
11	62.010	62.013	41.970	41.976	30.019	30.022
12	62.009	62.012	41.976	41.980	30.020	30.023
13	62.011	62.012	41.977	41.978	30.021	30.024
14	62.010	62.013	41.977	41.979	30.022	30.024
15	62.009	62.012	41.978	41.980	30.021	30.025
16	62.010	62.012	41.975	41.977	30.022	30.023
17	62.009	62.011	41.977	41.980	30.020	30.024
18	62.010	62.012	41.978	41.982	30.018	30.022
19	62.010	62.011	41.976	41.977	30.020	30.021
20	62.010	62.014	41.974	41.976	30.021	30.022
21	62.013	62.014	41.976	41.978	30.024	30.025
22	62.012	62.017	41.977	41.979	30.019	30.025
23	62.012	62.014	41.975	41.978	30.020	30.021
24	62.013	62.016	41.976	41.979	30.021	30.025
25	62.008	62.011	41.976	41.980	30.017	30.023

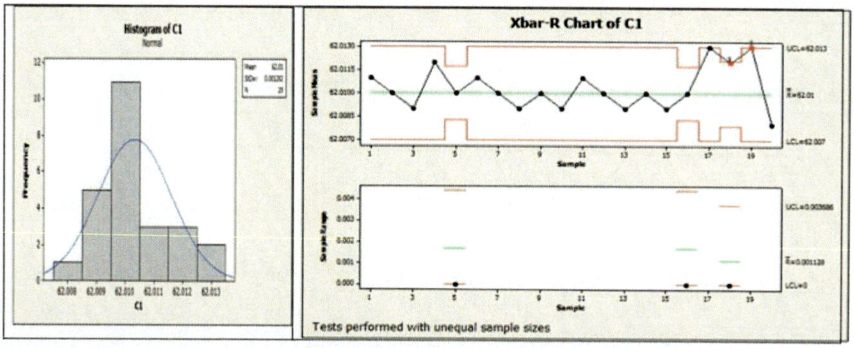

Fig. B.3 Histogram and X bar-R chart for part C1

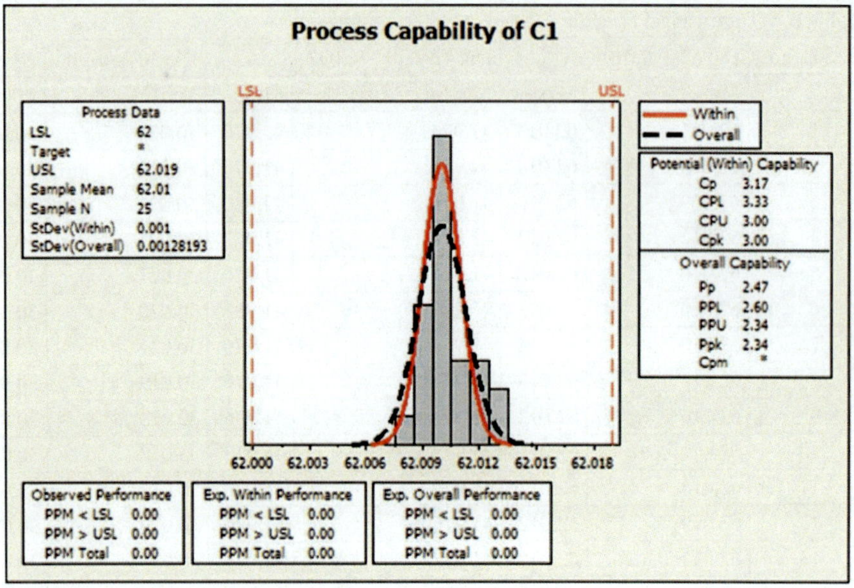

Fig. B.4 Histogram and X bar-R chart for part C1

obtained are presented in Table B.5 also presents the results of C_p, C_{pk} for other two dimensions Dia 42–0.017/−0.037 and Dia 30+0.038, respectively. Similar analysis has also been done for other two shifts, i.e., shift-II and shift-III. Tables B.6 and B.7 present the results.

From the results presented in Tables B.5, B.6, and B.7, following observations were made:

1. As evident from the results shown in Tables B.5, B.6, and B.7, that some of the C_p and C_{pk} values are >1 and some values are less than 1. If the value of $C_p < 1$, then process is not adequate. If the value of $C_p \geq 1.33$, then process is satisfactory enough. If the value is $1 < C_{pk} \leq 1.33$, then process is adequate.

Table B.5 Calculation table (Data set-I of diameter (mm))

Average	62.010	62.013	41.975	41.978	30.020	30.023
Max. size	62.013	62.017	41.978	41.982	30.024	30.025
Min. size	62.008	62.011	41.970	41.975	30.017	30.021
Range	0.005	0.006	0.008	0.007	0.007	0.004
s	0.001	0.001	0.002	0.002	0.002	0.001
C_{pk}	3.00	2.00	1.33	1.26	1.50	1.46
C_p	3.17	3.17	1.67	1.68	1.70	1.59

Table B.6 Calculation table (Data set-II of diameter (mm))

Average	21.518	21.521	31.974	31.976	34.974	34.978
Max. size	21.524	21.526	31.977	31.979	34.977	34.981
Min. size	21.514	21.517	31.970	31.974	34.970	34.975
Range	0.010	0.009	0.007	0.005	0.007	0.006
s	0.003	0.002	0.002	0.001	0.002	0.001
C_{pk}	0.73	2.88	0.91	2.53	3.77	3.09
C_p	0.73	2.88	0.91	2.53	3.77	3.09

Table B.7 Calculation table (Data set-III of diameter (mm))

Average	16.010	16.012	28.025	28.028	19.978	19.981
Max. size	16.013	16.014	28.030	28.032	19.986	19.990
Min. size	16.008	16.010	28.021	28.022	19.972	19.977
Range	0.005	0.004	0.009	0.010	0.014	0.013
s	0.001	0.001	0.002	0.002	0.004	0.003
C_{pk}	1.88	3.42	3.22	1.58	0.43	0.58
C_p	1.88	3.42	3.22	1.58	0.43	0.58

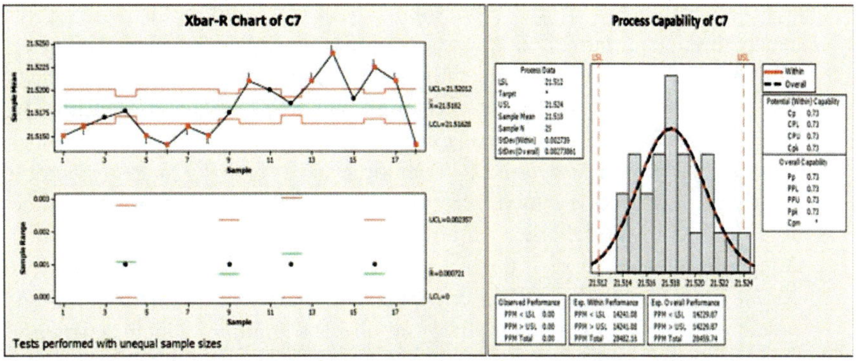

Fig. B.5 **a** X bar R chart for C7 of data set-2. **b** Process capability plot for C7 dimensions

2. With reference to data set 1 part dimension 1, values of *CP* and C_{pk} so obtained are >1.33 so, the process is assumed to be very satisfactory and we need not to change any parameters.
3. From Table B.7 data set -II, the part C7 with diameter 21.5 + 0.033 mm has most of the samples out of control as shown in the X bar and R chart (Fig. B.5a). So, the process capability values C_P and C_{pk}, so obtained are 0.73 and 0.73, respectively, which shows that the process is not adequate (Fig. B.5b). So, we need to change the parameters. Similar, results are observed from Table B.7. Data set-III where, part C17 and C18 have process capability values C_P and C_{pk} as

0.43 and 0.58, respectively, which shows that the process is not adequate to meet the requirements and needs further investigation.

4. Also, the influence of the measuring equipment/strategy (sampling size) and environment resulting in variability cannot be ruled out. All such possible causes that influence machining accuracy needs to be investigated. It is observed that even when the process becomes inadequate, i.e., obtained values are outside the specifications, still the operator continues with the operation/production, therefore institution of appropriate training methods for process control, need to be introduced.

4. Implications from Study

In nutshell following are the managerial implications of the study. Embarking upon them may help the organizations to successfully implement SPC program and reap the potential benefits.

(i) ***Management commitment and involvement***: Well-known quality gurus Deming, Juran, and Crosby stressed the need of top management commitment and involvement as one of the key dimensions for success of quality management initiatives in an organization (Deming 1986; Juran 1989; Crosby 1989). According to Robinson et al. (2000), until it becomes apparent to all employees that the SPC project is important to the management, it is very difficult to achieve coordinated efforts. From the study, it is learnt that the action about management commitment is to garner consistent support for quality improvement programs and follow management principles such as delegation of tasks, responsibilities and authority to the lowest possible level for successful SPC implementation thereby increasing efficiency.

(ii) ***Training and Education for SPC***: Training sessions for SPC awareness should be organized at regular intervals as the method is based upon statistical procedures and its implementation requires a good knowledge of selecting, calculating and interpretation of control charts (Antony 2000; Mahanti and Evans 2012). From the study, it is inferred that the main contexts of training and education for SPC include emphasis on on-hand training, which make engineers/managers well versed with data collection, analysis and design procedures of control charts. Experience in handling statistics provides the key for successful application of SPC.

(iii) ***Use of pilot projects and SPC facilitator***: To acquire an appreciation of the use of SPC as a powerful problem-solving tool it is essential to institute pilot projects with the help of SPC facilitator who will act as a "coach." Once SPC has been used successfully in one project/process, it is then easier to extend its use to other areas in an organization (Antony 2000). It is not practical to apply SPC manufacturing wide in the first instance due to cost and time constraints. For instance, in the study to discover and analyze the causes of variation resulting in rejection of crankcases, a pilot project to check for machining inaccuracy of crankcases was initiated. After formation of PAT along with SPC facilitator, the necessary data related to problem were collected and analyzed. The statistical parameters (average, range, etc.) were calculated to enable further process capability analysis and construction of the control charts.

(iv) ***Teamwork***: For the effective implementation and company-wide consciousness of SPC, it is strongly suggested to create cross-functional teams (CFTs) comprising of top management representative, steering committee and a process action groups/teams with operators, supervisor(s), process engineer(s), quality engineer(s), and an SPC expert/facilitator (Does et al. 1997). In the study for effective implementation of SPC and better understanding of the manufacturing process, a cross-function team with members from all departments (purchase, production and quality control) was constituted with an aim to coordinate tasks smoothly. The team carried out brainstorming sessions to note down causes of inherent variation and possible out of control situations.

(v) ***Choice of Control charts and measurement framework***: The selection of appropriate type of control charts and performance/process characteristic is absolutely vital for the success of any SPC program. Further, the measurement framework provides for the identification, definition, collection and analysis of performance measures that are used to understand, evaluate, predict, or control process variability Does et al. (1997). The type and selection of control chart can vary depending on characteristics of the process and the product to be controlled. A *control chart for variables* is used to monitor characteristics that can be measured and have a continuum of values, such as height, weight, or volume etc. A *control chart for attributes,* on the other hand, is used to monitor characteristics that have discrete values and can be counted. In the study, X bar and R charts are used to measure and analyze the shift data pertaining to machining of crankcases. Further, in order to know the inherent capability of the process, process capability plots are obtained. The use of measurement framework discussed in phase III of the proposed conceptual framework provided adequate understanding to interpret the results about the process.

5. Conclusion

The main aim of the study was to contribute significantly to the existing literature on quality control and quality improvement initiatives being undertaken by SMEs to improve the quality of their products/processes. Authors developed a conceptual framework, which connects theory with straightforward application of statistical process control in discovering and analyzing causes of variation to eliminate quality problems. The application of statistical tools such as Control charts, Histogram, Cause and Effect diagram along with Process capability analysis is presented in the study to eliminate quality problems arising out of various assignable causes during machining of crankcase. It is observed from the study that when the special or assignable causes of variation are present in manufacturing, the process is deemed to be out of control then use of SPC, as process monitoring methodology plays pivotal role in identification of these special causes of variation and helps to signal the need to take corrective action wherever and whenever appropriate. Disturbances resulting from manpower, machine, method/tooling and materials and other inaccuracies can be easily identified from the statistical process control. The process capability analysis and control charts are very simple to use provided possession of adequate knowledge and skills to use and interpret the results about the process.

Thus, from the study authors conclude that thorough understanding of the concept of variation, identification of causes of variation, and handling of these causes are important factors that can help the SMEs not only to improve their processes. Understanding the relationships between results and determinants makes it possible to have periodic feedback on the measures used (e.g., process capability indices), performance results (e.g., decrease in ppm) and to undertake incremental changes. This also supports the study conducted by Garengo et al. (2005) and Kumar et al. (2006) in which they emphasized on application of different quality programs on incremental manner rather than on radical manner in order to reduce the operational inefficiencies. They also advocated top management involvement and commitment in order to provide appropriate resources and training for successful implementation of quality initiatives.

Suggested Readings

1. Elg M, Olsson J, Dahlgaard JJ (2008) Implementing statistical process control: an organizational perspective. Int J Qual Reliab Manag 25(6):545–560

2. Garengo P, Biazzo S, Bititci US (2005) Performance measurement systems in SMEs: a review for a research agenda. Int J Manag Rev 7(1):25–47

3. Gullu A, Motorcu AR (2003) Elimination of the quality problems encountered during mass production by using statistical quality control. Turk J Eng Environ Sci 27:83–93

4. Antony J, Bhattacharyya S (2010) Measuring organizational performance and organizational excellence of SMEs—Part 1: a conceptual framework. Measur Bus Excell 14(2):3–11

5. Kumar M, Antony J, Singh RK, Tiwari MK, Perry D (2006) Implementing the Lean Sigma framework in an Indian SME: a case study. Prod Plan Control 17(4):407–423

6. Sharma RK, Kumar D, Kumar P (2006) Manufacturing excellence through TPM implementation—a practical analysis. Indus Manag Data Syst 106(2):256–280

7. Vassilakis E, Besseris G (2010) The use of SPC tools for a preliminary assessment of an aero engines maintenance process and prioritization of aero engines faults. J Qual Maintenance Eng 16(1):5–22